AF701395

INTEGRATED PEST MANAGEMENT FOR

WALNUTS

UNIVERSITY OF CALIFORNIA
STATEWIDE INTEGRATED PEST MANAGEMENT PROJECT
DIVISION OF AGRICULTURAL SCIENCES
PUBLICATION 3270 $12.00

1982

PRECAUTIONS FOR USING PESTICIDES

Pesticides are poisonous and must be used with caution. READ THE LABEL BEFORE OPENING A PESTICIDE CONTAINER. Follow all label precautions and directions, including requirements for protective equipment. Use a pesticide only against pests specified on the label or in published University of California recommendations. Apply pesticides at the rates specified on the label or at lower rates if suggested in this publication. Laws, regulations, and information concerning pesticides change frequently, so be sure the publication you are using is up to date.

Legal Responsibility. The user is legally responsible for any damage due to misuse of pesticides. Responsibility extends to effects caused by drift, runoff, or residues.

Transportation. Do not ship or carry pesticides together with food or feed in a way that allows contamination of the edible items. Never transport pesticides in a closed passenger vehicle or in a closed cab.

Storage. Keep pesticides in original containers until used. Store them in a locked cabinet, building, or fenced area where they are not accessible to children, unauthorized persons, pets, or livestock. DO NOT store pesticides with foods, feed, fertilizers, or other materials that may become contaminated by the pesticides.

Container Disposal. Dispose of empty containers carefully. Never reuse them. Make sure empty containers are not accessible to children or animals. Never dispose of containers where they may contaminate water supplies or natural waterways. Consult your county agricultural commissioner for correct procedures for handling and disposal of large quantities of empty containers.

Protection of Non-Pest Animals and Plants. Many pesticides are toxic to useful or desirable animals, including honeybees, natural enemies, fish, domestic animals, and birds. Crops and other plants may also be damaged by misapplied pesticides. Take precautions to protect non-pest species from direct exposure to pesticides and from contamination due to drift, runoff, or residues. Certain rodenticides may pose a special hazard to animals that eat poisoned rodents.

Posting Treated Fields. For some materials, re-entry intervals are established to protect field workers. Keep workers out of the field for the required time after application and, when required by regulations, post the treated areas with signs indicating the safe re-entry date.

Harvest Intervals. Some materials or rates cannot be used in certain crops within a specified time before harvest. Follow pesticide label instructions and allow the required time between application and harvest.

Permit Requirements. Many pesticides require a permit from the county agricultural commissioner before possession or use. When such materials are recommended in this publication, they are marked with an asterisk (*).

Processed Crops: Some processors will not accept a crop treated with certain chemicals. If your crop is going to a processor, be sure to check with the processor before applying a pesticide.

Crop Injury. Certain chemicals may cause injury to crops (phytotoxicity) under certain conditions. Always consult the label for limitations. Before applying any pesticide, take into account the stage of plant development, the soil type and condition, the temperature, moisture, and wind. Injury may also result from the use of incompatible materials.

Personal Safety. Follow label directions carefully. Avoid splashing, spilling, leaks, spray drift, and contamination of clothing. NEVER eat, smoke, drink, chew while using pesticides. Provide for emergency medical care IN ADVANCE as required by regulation.

ORDERING PROCEDURE

Write to: Agricultural Sciences Publications
Division of Agricultural Sciences
University of California
Berkeley, California 94720

Include publication number 3270.

Within the United States: Make check or money order payable to the Regents of the University of California. Price includes postage and handling charges. California residents must include sales tax or furnish a resale number. Orders must be accompanied by payment or a formal purchase order.

Outside the United States: Price does NOT include postage, handling. Please do not send payment; request a Pro Forma Invoice, which will include postage and handling. Indicate the number of copies desired and whether they should be shipped by surface or air mail. Please be prepared to make your payment in U.S. dollars, through a bank located in the United States.

Note: A 20 percent discount is given on orders of 10 or more copies sent to one address. All sales are final; no publication may be returned for cash or credit.

ISBN: 0-931876-50-8
Library of Congress No. 81-69166

Printed in the United States of America

Contributors and Acknowledgments

This manual was produced under the auspices of the Statewide IPM Project, Ivan J. Thomason, Director; A.P. Gutierrez and J.M. Lyons, Associate Directors.

Prepared by the IPM Manual Group of the Statewide IPM Project, U.C. Davis

Mary Louise Flint, Director
Brunhilde Kobbe, Senior Writer
Jack Kelly Clark, Photographer (except where noted)

Technical Coordinators

William W. Barnett, Area IPM Specialist, Cooperative Extension, Fresno
Clyde L. Elmore, Weed Scientist, Cooperative Extension, U.C. Davis
Steven G. Sibbett, Pomology Farm Advisor, Cooperative Extension, Visalia
Beth Teviotdale, Plant Pathologist, Cooperative Extension, Parlier

Contributors

Entomology:	Martin M. Barnes, William W. Barnett, Dennis J. Culver, Clarence S. Davis, William H. Olson, Helmut Riedl, Wallace R. Schreader
Nematology:	Winfield H. Hart, Benjamin F. Lownsbery
Plant Pathology:	John M. Mircetich, William J. Moller, Beth Teviotdale
Pomology:	David E. Ramos, Steven G. Sibbett
Vertebrates:	Rex E. Marsh, Terrell P. Salmon
Weed Science:	Clyde L. Elmore, Bill B. Fischer

Special thanks to

T.M. Aldrich, V, Brown, W.R. Clark, W.W. Coates, G.L. Dickey, E. Fereres, L.B. Fitch, D.L. Flaherty, H.I. Forde, J. Fox, R. Goldfien, K.S. Hagen, J.L. Hatfield, L.C. Hendricks, D.M. Holmberg, T. Jones, A.H. Lange, A. Lansing, M.V. McKenry, E. Mulrean, E.C. Mussen, R.M. Norris, R. Nowierski, J.W. Osgood, V. Polito, W.O. Reil, R.E. Rice, B. Rudd, P. Rude, M.N. Schroth, G. Smith, M.W. Stimmann, R. Tennyson, I.J. Thomason, K. Uriu, C. Weakley, F. Zalom and the California Department of Food and Agriculture for comments, review or help with photographs.

Information in this manual has been derived from research supported by the University of California, USDA Cooperative Extension IPM funds, the Walnut Control Board and members of the walnut industry.

Production

Editing: Mary Jean Haley, Laura Baker
Drawings: Sargina Tamimi
Design and Production Coordination: Naomi Schiff

Table of Contents

Common and Scientific Names of Pests

Vertebrates

California ground squirrel *Spermophilus beecheyi*
Eastern fox squirrel *Sciurus niger*
Eastern gray squirrel *Sciurus carolinensis*
Fox squirrel *Sciurus niger*
Gray squirrel *Sciurus carolinensis*
Pocket gopher *Thomomys* spp.

Sciurus carolinensis (Eastern) gray squirrel
Sciurus niger (Eastern) fox squirrel
Spermophilus beecheyi California ground squirrel
Thomomys spp. Pocket gopher

Insects

Buckeye scale *Diaspidiotus aesculi*
Calico scale *Lecanium cerasorum*
Codling moth *Laspeyresia pomonella*
Dusky-veined aphid *Callaphis juglandis*
European earwig *Forficula auricularia*
European fruit lecanium *Lecanium corni*
European red mite *Panonychus ulmi*
Fall webworm *Hyphantria cunea*
False chinch bug *Nysius raphanus*
Fig scale *Lepidosaphes conchiformis*
Frosted scale *Lecanium pruinosum*
Italian pear scale *Epidiaspis leperii*
Navel orangeworm *Amyelois transitella*
Olive scale *Parlatoria oleae*
Oystershell scale *Lepidosaphes ulmi*
Pacific flatheaded borer *Chrysobothris mali*
Pacific spider mite *Tetranychus pacificus*
Putnam scale *Diaspidiotus ancylus*
Redhumped caterpillar *Schizura concinna*
San Jose scale *Quadraspidiotus perniciosus*
Southern fire ant *Solenopsis xyloni*
Two-spotted spider mite *Tetranychus urticae*
Walnut aphid *Chromaphis juglandicola*
Walnut husk fly *Rhagoletis completa*
Walnut scale *Quadraspidiotus juglansregiae*

Amyelois transitella Navel orangeworm
Callaphis juglandis Dusky-veined aphid
Chromaphis juglandicola Walnut aphid
Chrysobothris mali Pacific flatheaded borer
Diaspidiotus aesculi Buckeye scale
Diaspidiotus ancylus Putnam scale
Epidiaspis leperii Italian pear scale
Forficula auricularia European earwig
Hyphantria cunea Fall webworm
Laspeyresia pomonella Codling moth
Lecanium cerasorum Calico scale
Lecanium corni European fruit lecanium
Lecanium pruinosum Frosted scale
Lepidosaphes conchiformis Fig scale
Lepidosaphes ulmi Oystershell scale
Nysius raphanus False chinch bug
Panonychus ulmi European red mite
Parlatoria oleae Olive scale
Quadraspidiotus juglansregiae Walnut scale
Quadraspidiotus perniciosus San Jose scale
Rhagoletis completa Walnut husk fly
Schizura concinna Redhumped caterpillar
Solenopsis xyloni Southern fire ant
Tetranychus pacificus Pacific spider mite
Tetranychus urticae Two-spotted spider mite

Diseases

Armillaria root and crown root *Armillaria mellea*
Blackline: Cherry leaf roll virus
Branch wilt *Hendersonula toruloidea*
Crown gall *Agrobacterium tumefaciens*
Deep bark canker *Erwinia rubrifaciens*
Phytophthora root and crown rot *Phytophthora* spp.
Shallow bark canker *Erwinia nigrifluens*
Walnut blight *Xanthomonas campestris* pv *juglandis*

Agrobacterium tumefaciens Crown gall
Armillaria mellea Armillaria root and crown rot
Cherry leaf roll virus: Blackline
Erwinia nigrifluens Shallow bark canker
Erwinia rubrifaciens Deep bark canker
Hendersonula toruloidea Branch wilt
Phytophthora spp. Phytophthora root and crown rot
Xanthomonas campestris pv *juglandis* Walnut blight

Nematodes

Ring nematode *Macroposthonia xenoplax*
Root-knot nematode *Meloidogyne* spp.
Root-lesion nematode *Pratylenchus vulvus*

Macroposthonia xenoplax Ring nematode
Meloidogyne spp. Root-knot nematode
Pratylenchus vulvus Root-lesion nematode

Weeds

Annual bluegrass *Poa annua*
Barnyardgrass *Echinochloa crus-galli*
Bearded sprangletop *Leptochloa fascicularis*
Bermudagrass *Cynodon dactylon*
Cheeseweed *Malva parviflora*
Common groundsel *Senecio vulgaris*
Common knotweed *Polygonum aviculare*
Common lambsquarters *Chenopodium album*
Common purslane *Portulaca oleracea*
Curly dock *Rumex crispus*
Dallisgrass *Paspalum dilatatum*
Field bindweed *Convolvulus arvensis*
Johnsongrass *Sorghum halepense*
Junglerice *Echinochloa colorum*
Large crabgrass *Digitaria sanguinalis*
Little mallow *Malva parviflora*
Morningglory *Convolvulus arvensis*
Mustards *Brassica* spp.
Nettleleaf goosefoot *Chenopodium murale*
Prostrate knotweed *Polygonum aviculare*
Puncturevine *Tribulus terrestris*
Redroot pigweed *Amaranthus retroflexus*
Tumble pigweed *Amaranthus albus*
Watergrass *Echinochloa crus-galli*
Wild barley *Hordeum leporinum*
Wild oats *Avena fatua*

Amaranthus albus Tumble pigweed
Amaranthus retroflexus Redroot pigweed
Avena fatua Wild oats
Brassica spp. Mustards
Chenopodium album Common lambsquarters
Chenopodium murale Nettleleaf goosefoot
Convolvulus arvensis Field bindweed, morningglory
Digitaria sanguinalis Large crabgrass
Echinochloa colorum Junglerice
Echinochloa crus-galli Barnyardgrass, watergrass
Hordeum leporinum Wild barley
Leptochloa fascicularis Bearded sprangletop
Malva parviflora Little mallow, cheeseweed
Paspalum dilatatum Dallisgrass
Poa annua Annual bluegrass
Polygonum aviculare Prostrate knotweed, common knotweed
Portulaca oleracea Common purslane
Rumex crispus Curly dock
Senecio vulgaris Common groundsel, ragwort
Sorghum halepense Johnsongrass
Tribulus terrestris Puncturevine

Integrated Pest Management for Walnuts

California produces about 95% of the English or Persian walnuts grown in the United States. Within the state, the 205,000 acres of walnuts grow mainly in six areas. These areas fall within two major climatic regions: the hot interior valleys and the cool coastal districts (Figure 1). The differences in climate determine to a large degree the choice of walnut cultivar; cultivar choice, in turn, influences pest problems and crop management options. In addition, the orchard site, soil type, choice of rootstock and the nature of the walnut tree itself have to be considered when making management decisions.

Walnut trees provide a favorable environment for a variety of organisms. Fortunately, only a few of these are potential pests. Their impact depends on their interaction with other organisms, the walnut tree, the orchard environment, and on the effectiveness of management actions taken against them. The better we understand these relationships, the better we can control pests or even prevent pest outbreaks.

Integrated pest management (IPM) provides a long-term management strategy for minimizing losses caused by pests with as little cost to the grower and disruption of the environment as possible.

Use this book to plan an IPM strategy for your orchard. The introductory chapters on walnut development, growth requirements, general management practices, and monitoring tools provide the background upon which the management guidelines in the pest sections are based. The Key to Damage Symptoms will help you recognize common problems in the orchard. The introductory parts of the insect, weed, disease and vertebrate chapters will tell you where and when major pests occur. Detailed descriptions and photographs of pests and damage are presented later in each chapter. These sections also discuss how you can enhance natural control factors, design a monitoring program and use control actions most effectively.

IPM is a flexible, evolving strategy that will be updated periodically as new information becomes available. Check regularly with your farm advisor for new developments. Because pesticide registrations change frequently, no specific recommendations are given here.

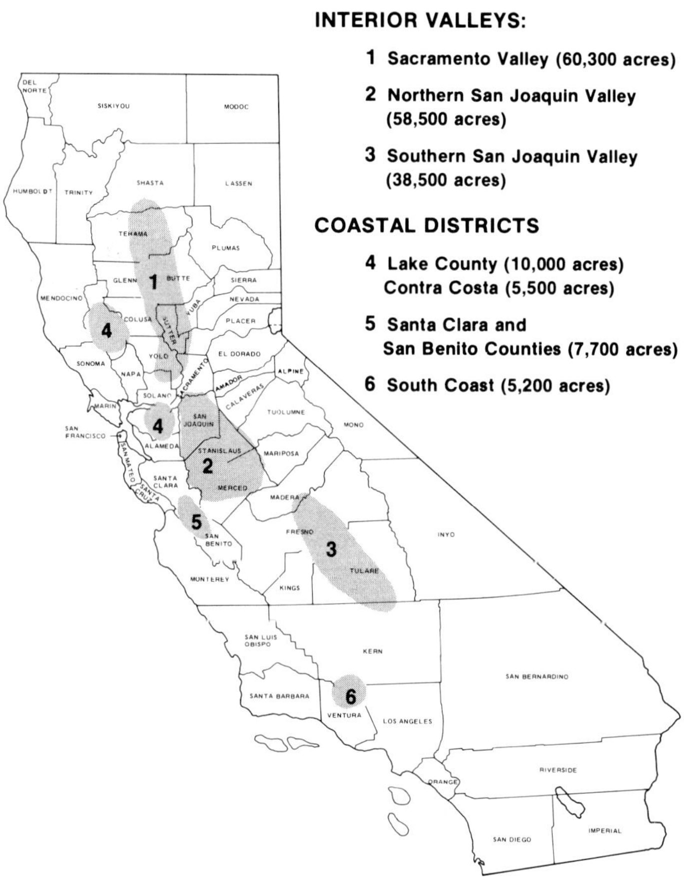

FIGURE 1. **Walnut-growing areas in California (total acreage, 1979)**

To obtain the latest information on pest control tools, see the current guides on insect, disease and weed control in walnuts, published by the University of California Division of Agricultural Sciences.

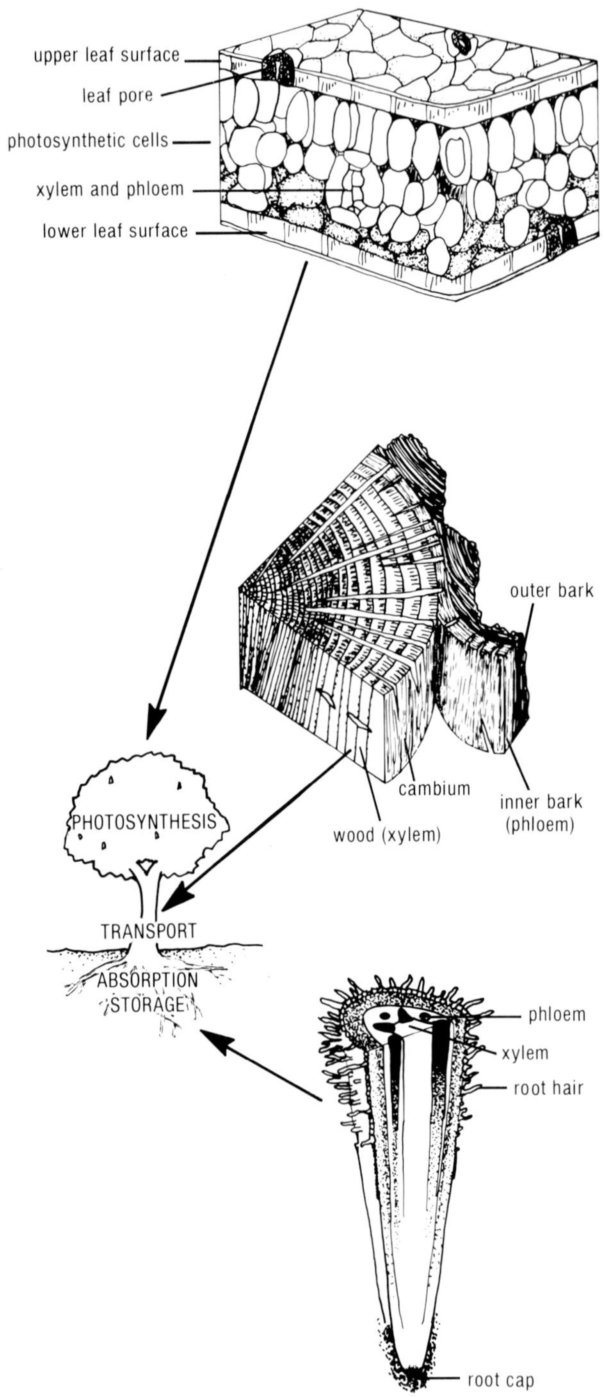

FIGURE 2. Basic tree structures and functions. The young roots absorb water and minerals, which are transported in the xylem to the canopy, where leaves produce sugars in the process of photosynthesis. The tree uses the sugars immediately for growth or transports them in the phloem mainly to the roots for storage.

The Walnut Tree: Development and Growth Requirements

The objective of walnut production is to harvest a large crop of light-colored, plump kernels. Optimal production can be achieved only with healthy trees. Maintaining tree health requires a basic understanding of the seasonal cycle, the crop's specific requirements and the impact that pest organisms and cultural or environmental factors have on tree and nut growth. Without this understanding, it is easy to overlook stress symptoms caused by environmental factors or to confuse them with pest damage. An even more important aspect of IPM is the evaluation of cultural practices and pest control methods for their impact on the total orchard system as well as for their effectiveness in eliminating the primary pest problem. In some cases, well-intended practices compound stress in the crop and add to economic loss.

The Seasonal Cycle of the Walnut

The walnut tree spends the winter in a dormant state. During dormancy, the tree does not grow but maintains a base level of water transport and starch consumption. The breakdown of starch increases the sugar concentration in the cell sap, preventing it from freezing. To produce enough starch for this process, the tree must have been provided with sufficient water in the fall.

Early in spring, when temperatures become favorable, stored nutrients are mobilized and used by the buds, providing the energy to develop the flowers and leaves. Once green leaves begin to form, they provide the energy for further growth through photosynthesis. Figure 2 shows the basic tree structures responsible for storage, transport and energy production.

Walnuts produce male and female flowers on the same tree. Both develop in buds formed during the previous season. The female or pistillate flowers are hidden inside leaf buds at the tip of the previous year's terminal shoots; in some cultivars, lateral shoots also bear pistillate flowers. The male flowers or catkins are visible above the leaf scars on the previous season's shoots. In most cultivars, the first sign of growth in the spring is the enlarging of the catkins; then the leaf buds burst with their first flush of shoot growth, and the pistillate flowers begin to develop (see photos on back cover).

Walnut trees leaf out from early March to the middle of April, depending on the cultivar and the climate. Among the commonly planted cultivars, Payne is one of the first cultivars to leaf out in the spring. All cultivars are ranked as early, mid-season, or late in reference to Payne. The time of leafing out is important because it usually determines the timing of later developmental stages and tree or nut susceptibility to damage by various environmental stresses and certain pests.

Shoots and leaves grow rapidly, and the foliage is partially expanded by the time the female bloom begins. A second flush of shoot growth occurs during summer. The absence of this second growth flush is a good indicator of poor growing conditions.

The pistillate flowers form at the tips of the first flush of spring growth. The pollen is carried from the catkins to the pistillate flowers mostly by the wind. Although the pistillate bloom extends over about a month, the peak pistillate receptivity in many cultivars occurs before or after the main pollen-shedding period, making it difficult for the catkins to pollinate all female flowers of the same cultivar. To enhance pollination, the orchard is often interplanted with about 10% of a cultivar that sheds its pollen during peak pistillate bloom of the major cultivar.

After fertilization, the female flower begins to develop into the nut. First, its size increases; good nut growth depends on having ample soil moisture available at this time. By about seven weeks after bloom, the husk and shell have reached almost full size and the shell begins to harden (Figure 3). Next season's catkin buds form before shell hardening; pistillate flower buds form some time after shell hardening, depending on the cultivar.

About 15 weeks after bloom, the tissue inside the shell begins to form the kernel or nutmeat. The dry weight and the oil content of the kernel continue to increase through harvest. Stresses caused by pests, environmental factors, or lack of water at any time during this process will cause significant loss in nut quality and yield. At maturity, the packing tissue around the kernel turns brown and the kernel quality is at its highest point. Then the husks begin to split, exposing the nuts to invasion by the navel orangeworm and molds.

To preserve quality and prevent damage, harvest the nuts as soon as at least 80% of them can be shaken from the tree and 95% are hullable. Figure 4 shows a mature, dried kernel.

During the late summer, sugar produced in the leaves will be transported to buds, limbs, and roots, where it is stored for the next season's initial growth. The seasonal cycle is completed when, in the late fall, the tree prepares for dormancy by moving nutrients from the leaves into twigs and stems, reducing its moisture content, converting starch to soluble carbohydrates and shedding the leaves.

FIGURE 3. At about seven weeks after bloom, the nuts have reached full size and the shell begins to harden. The kernel starts to form, replacing the liquid tissue containing stored food.

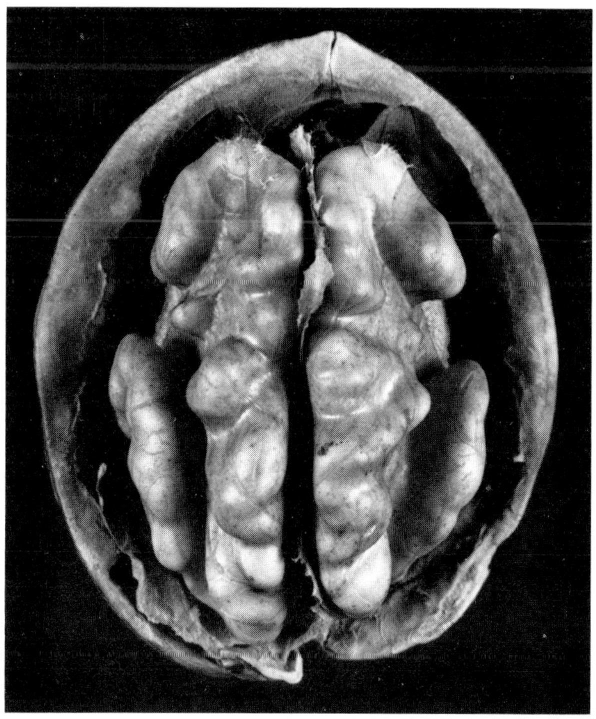

FIGURE 4. A high quality, mature kernel has plump, light nutmeat and a light brown membrane. The kernel halves are separated by the packing tissue.

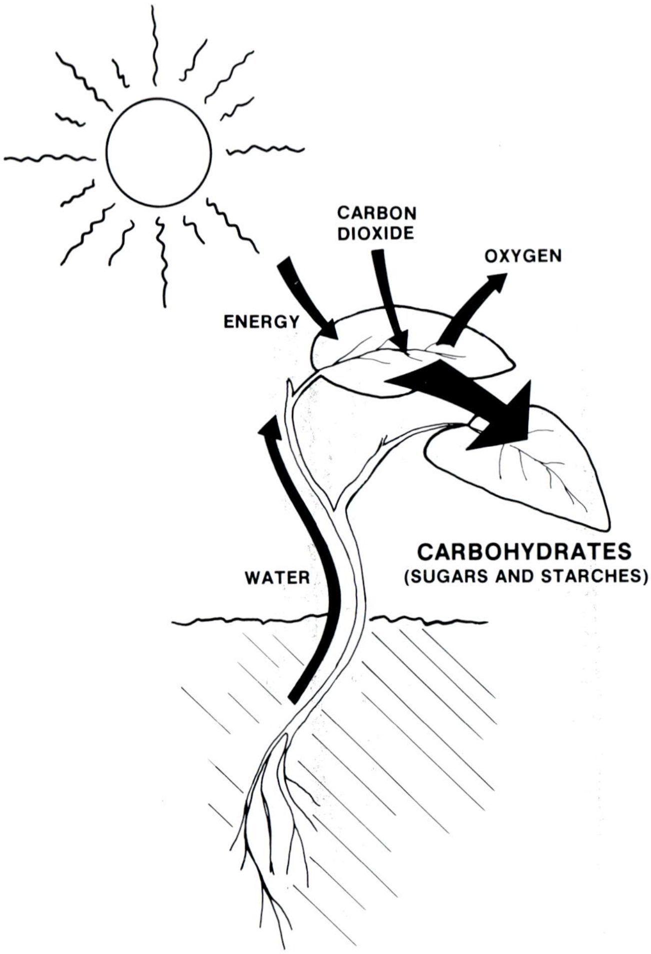

FIGURE 5. In photosynthesis, green plants use energy from the sun to convert carbon dioxide from the air and water from the soil into sugars, their primary food source. In the process, oxygen is released into the air.

Growth Requirements

Tree growth and nut development require sufficient light, water, and nutrients and adequate temperatures. Light is essential for photosynthesis, the process by which green plants manufacture sugars, their major food source (Figure 5). The sugars are synthesized in the leaves from water taken from the soil, carbon dioxide taken from the air, and light energy taken from the sun. Sugar provides the energy for tree growth, development, and maintenance and may be used immediately or stored as starch in leaves, stems, or roots.

Irrigation is necessary for profitable walnut production in the semi-arid climate of California. In the Central Valley, a mature walnut orchard requires 48 to 54 acre-inches of water per year for optimal yield of high quality nuts. Only a small fraction of the water taken up by the tree is retained; most of it passes through the plant and out the leaf pores during the process called transpiration. Transpiration is necessary to supply water to the leaves for photosynthesis, to carry nutrients to all parts of the tree, and to cool the plant. Climatic factors, such as the amount of sunshine, humidity, heat and wind, affect the rate of transpiration.

Various nutrients are needed for good tree growth and nut production. California orchard soils contain sufficient amounts of most essential nutrients either because they are already present in the soil or are returned periodically through leaf decay. Others, nitrogen in particular, have to be replenished through fertilization.

Walnut trees grow best under mild temperatures but require a certain amount of winter chilling for uniform bloom and good fruit set in the spring, especially in late-leafing cultivars. Walnuts are susceptible to extreme temperatures, which can cause frost or sunburn damage (see page 72).

Managing Pests in Walnuts

Integrated pest management is based on an understanding of the factors that make a walnut orchard vulnerable to pest invasion or damage. The best management strategy not only controls pests but anticipates and prevents problems. Integrated pest management can enhance your current control program; it also offers guidelines for selecting optimum control methods. Minimizing pest-caused crop losses and environmental damage and optimizing crop production are the ultimate goals of IPM.

An integrated pest management program has four major components:

- pest identification
- field monitoring
- control action guidelines
- effective methods for prevention and control

Pest Identification

Correctly identifying pest species and stress symptoms is fundamental to choosing corrective management strategies. Most pest management tools, including pesticides, are effective only against a select group of species; therefore, different practices may be needed even for closely related species. In some cases, the symptoms caused by pest organisms closely resemble those caused by nutrient deficiencies or other soil problems. The Key to Damage Symptoms, the photographs, and the seasonal charts in this manual are intended to help in identifying problems. Farm advisors can assist you in confirming identification or in identifying unusual problems.

Field Monitoring

Field monitoring provides information on daily or seasonal orchard conditions, such as the status of pests, weather, or soil moisture. This information can be used to evaluate and predict potential pest problems. Because conditions vary even between neighboring orchards, every orchard must be monitored. Check regularly the pest species present, the maturity and health of the crop, the weather, plant environment and, when appropriate, the population levels of pest and beneficial organisms. Keep records of monitoring results, weather and management activities. During the season, these records can show whether pests or natural enemy populations are rising or decreasing. They can also help forecast possible outbreaks of the next generation of a pest. Pest population counts, together with records of control measures, cultural practices and weather conditions, can help decide whether future control actions will be needed. Simple tables and graphs of data help define patterns, and maps help identify localized problems and pest distribution. Over the years, these records provide valuable historical data for long term orchard management.

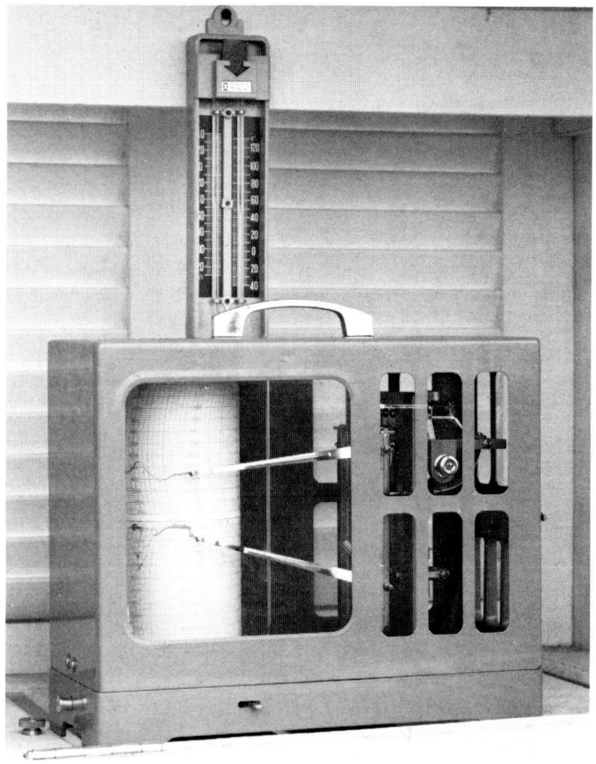

FIGURE 6. A maximum-minimum thermometer (top) records the highest and lowest temperatures over a 24-hour period. A hygrothermograph (below) records temperature and humidity continuously over several days.

Monitoring Pests

The frequency of monitoring varies with the individual pest species and the season. Sampling programs for most species are needed only at certain times of the year. Start sampling well before populations begin to build and continue sampling through the pest's damaging stages. Monitor every orchard for insect pests and

FIGURE 7. Protect your instruments for measuring temperature and humidity in a vented weather shelter.

their natural enemies at least once a week when pests are likely to be present and more frequently when insect populations can be expected to flare up.

No weekly sampling programs have been developed to assess populations of weeds, diseases, nematodes and vertebrates. It is important, however, to note their presence and watch for changes in their status. Prepare a record once each quarter (fall, winter, spring, and summer) for each orchard, noting the dominant weeds, diseases, vertebrates and any other significant stresses. Correct identification is essential. Sample for nematodes when you suspect they are causing losses or when the site history indicates they may become a problem. Specific monitoring procedures are described in the discussions of individual pests.

Monitoring Weather

Weather greatly influences the development of both the walnut tree and its pest species. Good weather information is necessary for timing cultural operations and pest control activities. Local newspapers and radio stations report daily high and low temperatures, relative humidity, wind conditions, and evapotranspiration data. The National Weather Service broadcasts local and regional weather information in California over two very high frequency (VHF) FM bands, 162.40 and 162.55 megahertz. The broadcasts include climatological data, frost warnings, agricultural weather forecasts and advisories, river conditions, as well as emergency warnings of hazardous conditions. The taped messages are repeated every 4 to 6 minutes 24 hours a day and revised every 2 to 3 hours.

Weather conditions, particularly temperature and rainfall, may vary considerably within a few miles due to differences in elevation, vegetation cover, or amount of water present. To have the most accurate data for your orchard, you may want to set up a basic weather station. A maximum/minimum thermometer or thermograph measures the highest and lowest temperature of the day (Figure 6). This information is used to calculate degree-days, which currently are used to forecast the development of a major pest, the codling moth, and may be used in other aspects of orchard management in the near future. A precipitation gauge, which measures the amount of rainfall, is important in setting up a water budget.

Instruments to measure evaporation (a Class A evaporation pan), humidity (a dry/wet bulb thermometer or a hygrothermograph), wind speed (anemometer), and wind direction (wind vane) are also available. Enclose the thermometer and hygrothermograph in a weather shelter (Figure 7). Maintain all instruments according to the manufacturer's instructions and calibrate them regularly to assure accuracy. Keep records of all your observations.

Accumulating Degree-Days

The growth rates of the plant and its pests are closely related to temperature. For instance, in a warm year both will develop faster than in a cold year. Therefore, the calendar date alone is not a precise guide for carrying out crop and pest management activities. For more accurate predictions of crop and pest development, time and temperature have to be considered together. The rate of development depends on the amount of heat the plant or pest experiences during a given time period. Measuring the amount of heat accumulating over time provides a 'physiological' time scale, which is biologically more meaningful and more accurate than calendar days. The unit of the physiological time scale is the degree-day (°D). It is defined as one degree above the developmental threshold maintained for a full day. You can measure the temperature in Fahrenheit or Celsius; all degree-day figures in this manual are based on the Fahrenheit unit.

Each plant and insect species has a specific lower and upper developmental threshold. The lower threshold is the temperature below which no development occurs. Above this threshold, the rate of development increases with temperature at a roughly linear rate until further growth is inhibited at the upper developmental threshold. An organism's lower and upper thresholds must both be known to calculate degree-days accurately.

To use degree-days for keeping track of crop or pest development, you must know the amount of heat (number of degree-days) required to complete each growth stage. At high temperatures, degree-days accumulate faster, reflecting faster development. For example, codling moth eggs do not develop below 50 °F and require 160 °D above this threshold to hatch. When 10 °D accumulate every day, eggs hatch after 16 days. Where 20 °D accumulate per day, the eggs hatch after only 8 days.

You can estimate degree-days from the daily minimum and maximum temperatures as outlined in Table 1. This formula provides an approximate guide and is useful only if the minimum temperature is above the lower threshold. More precise formulas have been worked out, and with the aid of a computer, a degree-day table can easily be compiled for an organism if the threshold temperatures are known.

Table 1. Sample of Degree-Day Calculation

Degree-Days (°D) accumulated for one day:	Average daily temperature	–	Lower developmental threshold	=	°D
Add daily minimum and maximum temperatures and divide by 2 to get the average daily temperature.			$\frac{74\ °F + 54\ °F}{2} = 64\ °F$		
Subtract the lower threshold temperature (e.g., 50 °F for the codling moth) from the average daily temperature. The result is the number of degree-days accumulated that day.			64 °F – 50 °F = 14 °D		

Control Action Guidelines

Control action guidelines indicate when to take management actions to avoid eventual crop losses. They are meaningful only in conjunction with careful field monitoring and accurate pest identification. For several insect pests, numerical thresholds have been developed that indicate economically damaging population levels, and these have been used as a basis for control action guidelines. However, for most diseases, weeds, nematodes, and vertebrates, the guidelines are currently less specific and may be based on combining information about the extent of pest infestation, the potential for effective and economical control, the history of an orchard or a region, the stage of crop development, or the weather.

Management Methods

The best methods for an IPM program are those that prevent pest outbreaks and provide long-term and economical control. Preventive methods include choosing cultivars and rootstocks less susceptible to pest damage, removing overwintering sites or sources of new infection, choosing the appropriate irrigation and cultivation method, and timing the various management activities to discourage pest development and keep trees growing vigorously. Use pesticides only when careful field monitoring indicates that they are needed to prevent economic loss.

Cultivar and Rootstock Selection

The selection of a cultivar and rootstock is possibly the most critical planning decision because it affects management practices, pest problems, and profits throughout the life of the orchard.

All scion cultivars are derived from the English walnut, *Juglans regia.* In the past, they were primarily selected for increased nut yield and quality. Current breeding programs are also selecting for a late leafing time to avoid early season pest problems and to spread the harvest season. The major cultivars currently available vary in their susceptibility to pest damage (Table 2). The potential for damage by some pests, such as blight and codling moth, is mainly due to the time of leafing and blooming; for others, such as the husk fly, chemical

Table 2. Early, Mid- and Late Season Scion Cultivars and Their Susceptibility to Major Pests

CULTIVARS[1]	SUSCEPTIBILITY[2] TO		
	Blight	Codling Moth	Walnut Husk Fly
Early			
Amigo	++	++	NE
Ashley	+++	+++	+
Chico	++	+++	++
Gustine	+++	++	NE
Marchetti	+++	++	NE
Payne	+++	+++	++
Serr	++	+++	NE
Sunland	+++	NE	NE
Mid-Season			
Chandler[3]	+	+	NE
Eureka	+	+	+++
Hartley	+	+	+++
Howard[3]	+	+	NE
Tehama	+	++	++
Vina	+	+	NE
Pedro	+	+	NE
Late			
Franquette	+	+	+++

[1] grouped in relation to leafing date of Payne

[2] +++ = very susceptible; ++ = fairly susceptible; + = fairly tolerant; NE = no experience

[3] new cultivars; experience limited

or morphological traits of the cultivar probably play a role.

Two rootstocks have been selected both for vigorous growth and for tolerance to pathogens and nematodes (Table 3). Seedlings of the Northern California black walnut (*Juglans hindsii*) are resistant to the root-knot nematode and are highly tolerant of *Armillaria*, but they are susceptible to other diseases and to certain other nematode species. Seedlings of the Paradox hybrid (*J. hindsii* x *J. regia*) tolerate heavy clay and clay loam soils and are generally more tolerant of adverse soil conditions and diseases. They are, however, more sensitive to soil salinity than black walnut. Seedlings of English walnut (*J. regia*) are not recommended as rootstock because they are susceptible to pathogens, nematodes, and salinity.

Cultural Practices

Ground Cover. Plants growing on the orchard floor may compete with trees for nutrients and water, contribute to pest problems, weaken young trees or interfere with sprinkler operation. However, a ground cover can be desirable in a mature orchard if it is properly managed.

A ground cover often improves water penetration and aeration of the soil, two factors important for a healthy root system. In addition, a well-managed ground cover provides cooling and reduces dust, which often contributes to mite outbreaks. The cooling effect of the ground cover can be advantageous in areas with hot summers but a disadvantage in areas with frost hazards.

Table 3. Susceptibility of Major Rootstocks to Pathogens, Nematodes and Soil Salinity

Rootstock	SUSCEPTIBILITY[1] TO						
	Blackline[2]	Armillaria root and crown rot	Phytophthora root and crown rot	Crown gall	Root knot nematodes	Root lesion nematodes	Salinity
Northern California black walnut (*Juglans hindsii*)	+++	+	+++	+	+	+++	+
Paradox (*J. hindsii x J. regia*)	+++	++	++[4]	+++	NE[5]	+++	++
English[3] (*J. regia*)	+	+++	+++	+++	+++	+++	+++

[1] +++ = susceptible; ++ = fairly tolerant; + = highly tolerant

[2] Black walnut and Paradox are immune to the virus, but English scions on them are killed by the disease. Although English scion on English rootstock does not develop the black line (girdling) at the union, infected trees may be carriers of the virus and are often stressed by other unknown factors

[3] *J. regia* is not recommended as a rootstock

[4] Paradox is significantly more resistant to *P. cactorum* and somewhat more resistant to other *Phythophthora* species than the other two rootstocks

[5] NE = no experience

A ground cover may also be important in providing food and shelter for many predators and parasites of insects and mites. More research on the interaction between different types of ground cover and beneficial and pest species is necessary to determine how vegetation management can be used to its full extent in an IPM program.

To maintain a ground cover properly, be sure to mow frequently enough to avoid trash buildups and to prevent weeds from seeding. Keep plant growth away from the tree trunks, where it provides conditions favorable for crown rot infections. An orchard with a ground cover uses more water and nutrients than a clean-cultivated orchard; the cost of adjusting irrigation and fertilization programs to account for this extra use must be considered. See the weed chapter for details of vegetation management methods.

Sanitation. Sanitation is an important management tool for reducing the buildup and spread of insects, pathogens and weeds in the orchard. Shake mummy nuts from the trees and remove and destroy the trash nuts to deprive the navel orangeworm of overwintering and breeding sites. Controlling lichens on trunks and limbs will remove shelter for the Italian pear scale. Obtain only pest-free nursery material and clean equipment brought in from another area to avoid introducing pathogens, insects, nematodes or weeds into the orchard. Keep tree crowns dry and free of weeds to reduce infections of *Phytophthora* fungi. If you find a very small infestation of a perennial weed in or around the orchard, eradicate it before it spreads throughout the orchard.

Water Management. The objective of water management is to supply the tree with the right amount of water at the right time so that the tree can make efficient use of nutrients from the soil and produce an optimum yield of quality nuts. Figure 8 shows the monthly water requirements for mature walnut trees in the Central Valley. Careful water management also helps prevent certain stress and pest problems. Water stress early in the spring leads to small or wilting leaves and small nuts. In a low rainfall year, a winter irrigation may be necessary to fill the root zone with water before growth begins. Severe water stress during the summer may cause leaf drop, sunburn on nuts and limbs, and sticktight nuts at harvest. The damaged nuts provide breeding sites for a major pest, the navel orangeworm, and sunburned limbs are susceptible to branch wilt disease. Overwatering can lead to waterlogging of the soil, contributing to root and crown rot diseases. Water quality is also important; frequently, irrigation water contains high levels of boron or chlorine, which cause toxic symptoms (see page 73). More information on water analysis and corrective measures is available at farm advisors' offices.

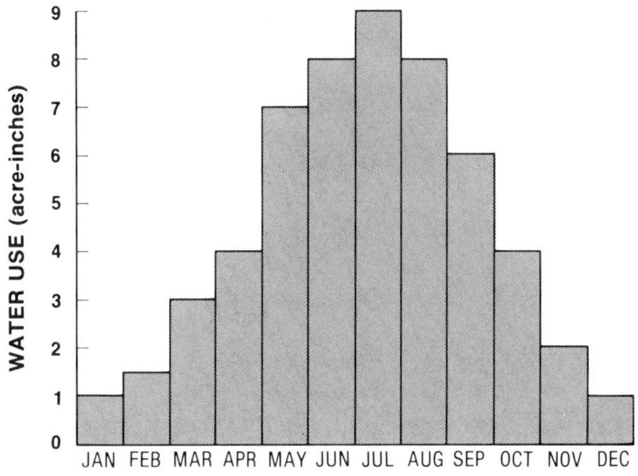

FIGURE 8. Average monthly water use by mature walnut trees with full canopy in the Central Valley.

FIGURE 9. In furrow irrigation, the water is channeled along furrows across the orchard.

FIGURE 10. With border irrigation, the total land area between borders is flooded. Border and furrow irrigation can only be used on flat land.

FIGURE 11. Sprinkler irrigation can be used on flat as well as rolling or sloping land.

Irrigation Methods. Several methods can be used to irrigate walnut orchards, including furrows, borders, sprinklers and low volume irrigation (Figures 9–11). Your choice will depend on availability of water, installation and operation costs, soil texture, slope of the land and frost hazard. Each system can supply water adequately if managed properly. The new low volume irrigation techniques, such as drip, mist and mini-sprinklers, are designed to meet the tree's water requirements on a daily basis. Experience with their use in walnuts is limited.

Certain pest problems can be affected by the irrigation method chosen. Border or furrow irrigation can spread weeds and pathogens through the orchard as the water moves across the surface. If the water comes from canals, it may be contaminated with weed seeds and pathogens. On the other hand, border and furrow irrigation discourages ground squirrels and pocket gophers from digging burrows in the orchard. Sprinklers can increase blight problems if water from high-angle sprinklers wets the leaves or nuts. Drip irrigation hinders weed control because it keeps the area around the emitters moist, facilitating rapid breakdown of preemergence herbicides.

Scheduling Irrigations. To avoid over- or underirrigating, carefully time irrigations and apply the correct amount of water. If you wait until your trees show signs of stress, such as drooping or wilting leaves, crop damage may already have occurred. As trees mature, adjust the amount of irrigation water to account for the increased demand. For details on irrigation scheduling, see UC Agricultural Science Publication 21199 or consult your farm advisor. A brief summary is given below.

You can determine when and how much water to apply by compiling a water budget and measuring soil moisture. First, you have to know the available water (AW), the amount of water that your orchard soil can store. Your farm advisor can provide you with AW levels for various soil types. Using AW and the depth of the trees' roots, you can calculate the total amount of water available to the tree when the soil is filled to field capacity. In many areas this water reservoir is filled early in the spring from normal winter rains. If the soil water is not up to capacity, a prebloom irrigation is needed.

To compile the water budget, you must determine how much water is being used in the orchard each day or week and subtract that amount from the water reservoir. When the soil moisture reaches a predetermined allowable depletion level, it is time for an irrigation to refill the reservoir. The allowable depletion depends on many factors, including soil type and developmental stage of the plant. For walnuts, it may range from 30% to 50% of the water reservoir. Your farm advisor or irrigation specialist can help you estimate the allowable depletion for your orchard. This amount of water plus an allowance for system inefficiency and leaching has to be applied at each irrigation.

You can estimate how much water is being used by keeping track of soil moisture with tensiometers, gypsum blocks or soil moisture cells. Generally, the instruments are installed at two depths, one in the upper half and one in the lower half of the root zone, and read twice a week. You can also estimate water loss by using evapotranspiration (ET) data and checking soil moisture periodically to adjust for irrigation inefficiency. ET is defined as the amount of water lost through transpiration by the plant and through evaporation from the soil. ET data for the current season are compiled by many farm advisors' offices and reported in bulletins and local papers. If current ET data are not available, historical data for a "normal" year can provide an approximation, especially in midsummer when the weather does not vary much from year to year. ET figures given for walnuts are based on orchards with over 80% canopy cover.

Several new developments promise to make water management more accurate and easier in the near future. An instrument called a neutron soil moisture probe has been developed to measure water depletion. There is also a portable pressure chamber that measures water stress on the tree directly, using the tree itself as a tensiometer. Investigations are underway to determine the allowable stress of walnut trees, as measured by the pressure chamber at predawn, during critical stages of crop development, such as pollination, fruit growth and kernel development. This system will allow you to schedule irrigation based on the tree's daily and seasonal needs.

Finally, the University, in cooperation with the California Department of Water Resources, is developing a centralized computer system that can accurately estimate irrigation needs in individual orchards based on crop ET requirements, weather, and specific soils information. Check with your farm advisor to find out if this program is available in your county.

Fertilization. Fertilization is usually necessary to supply the trees with adequate amounts of all essential nutrients. Trees weakened by nutrient deficiencies produce less and are more susceptible to pest damage than healthy trees. California orchard soils contain sufficient amounts of most essential nutrients; usually only nitrogen has to be replenished on a yearly basis. Occasionally, zinc, potassium, magnesium, manganese and copper are deficient; phosphorus is rarely deficient.

The tree's ability to take up nutrients also depends on the type of fertilizer, soil moisture, soil type and temperature. Make sure you apply the appropriate fertilizer at the right time and place it within reach of the root zone. If you maintain a ground cover, allow for the extra need in fertilizer.

Some nutrient deficiencies produce characteristic leaf symptoms (see page 73), but you should confirm suspected deficiencies with a leaf analysis performed by a professional laboratory. Check with your farm advisor or PCA for laboratories and recommended procedures in collecting leaf samples. Take leaf samples every year in June or July to detect early any deficiencies or possible toxicities. Table 4 lists the critical levels of those nutrients that are most often deficient or present at toxic levels.

Table 4. Critical Nutrient Levels in Walnut Leaves[1]

Nutrient	deficient below	toxic above
nitrogen	2.1%	
potassium	0.9%	
zinc	15 ppm	
boron	20 ppm	300 ppm
sodium		0.1%
chlorine		0.3%

[1] determined by leaf analysis performed in July; percent or parts per million (ppm) based on dry weight.

The severe lack of one nutrient may hamper the uptake of other nutrients. If the leaf analysis shows that several nutrients are deficient, first add the one that is most deficient. Retest the following summer and add other nutrients if necessary.

If a leaf analysis reveals toxic levels of minerals, such as sodium, boron or chlorine, a soil analysis should be performed. A soil analysis is useful in locating toxic levels of these minerals and in deciding what corrective measures to take.

Harvesting. Nuts are ready for harvest when the packing tissue has turned brown; at this point, kernel quality is highest, and any delay in harvest will cause some loss in quality. Harvest the nuts as soon as 80% of the nuts can be removed and 95% are hullable at the first shake. The remaining nuts can be harvested about 10 to 14 days later in the second shake. If harvest is delayed or if the nuts remain on the ground too long, the kernels may be damaged by high temperatures, molds, or the navel orangeworm, which will enter undamaged nuts after their husks have split. Prompt harvest, rapid drying and fumigation will preserve nut quality and prevent damage.

The kernel and husk mature at about the same time along the coast, but in the interior valleys the kernel matures two to three weeks ahead of the husk. Hull split

Table 5. Some Common Natural Enemies Attacking Walnut Pests

	PESTS COMMONLY ATTACKED				
	Aphids	Armored Scales	Soft Scales	Mites	Navel Orangeworm
PREDATORS					
convergent lady beetle	●				
ashy gray lady beetle	●				
twice-stabbed lady beetle		●			
spider mite destroyer				●	
Cybocephalus		●			
green lacewings	●			●	
minute pirate bug	●			●	
six-spotted thrips				●	
western predatory mite				●	
PARASITIC WASPS					
Trioxys	●				
Praon	●				
Aphytis		●			
Prospaltella		●			
Metaphycus			●		
Pentalitomastix					●

The walnut aphid parasite, *Trioxys pallidus*

can be promoted by spraying with a growth regulator, such as ethephon; this often enables you to harvest one to two weeks earlier than under natural maturation of the husk. Ethephon is fairly expensive, and application must be thorough and carefully timed. To gain experience with a growth regulator, first apply it on a small portion of the orchard. Consult your farm advisor or PCA for details.

Biological Control

Potential pest populations in the walnut orchard are frequently controlled by other organisms inhabiting the orchard. These biological control agents include predators, parasites, pathogenic microorganisms, and competitors. They may be native or introduced from other areas. The importance of biological control in walnuts is better known for insects and mites than for other pests, although no effective natural enemies have yet been found for such key insect pests as codling moth, navel orangeworm and husk fly. Table 5 lists some common natural enemies that attack walnut pests.

Biological control may be disrupted by chemical treatments for key pests, resulting in outbreaks of aphids, mites or scales. You can reduce such secondary outbreaks by carefully selecting types and rates of insecticides and timing applications so that they are least disruptive to the natural enemies.

Biological control of diseases and weeds is very limited at present. One walnut disease, crown gall, can be controlled on young trees by a competitor, a nonpathogenic strain of the crown gall pathogen. The introduction of the puncturevine stem weevil and seed weevil has helped control puncture vine in orchards.

Pesticides

Within an IPM program, pesticides are valuable pest management tools because they can reduce pest populations drastically within a short time. In many cases, they are the only control tools available.

Use of Pesticides. Pesticides, including insecticides, herbicides, fungicides and rodenticides, belong to a variety of chemical groups with very different properties. Often you can choose among several chemicals and different formulations to control one pest. On the other hand, you may be able to select one chemical to control several pests at the same time or use compatible materials together. The choice will depend on the degree of control needed, the effect on other pests and beneficial species, and on economic and legal restrictions. Where possible, choose the least disruptive chemical. Some synthetic organic materials used for control of the codling moth, for example, are more disruptive to natural enemies than others. However, they may be more effective and may be required when infestations are severe. Among the most selective materials available are preparations of disease-causing microorganisms. These microbial insecticides are desirable control tools in IPM programs because they act selectively against certain insect groups and are nontoxic to natural enemies and nontarget species. *Agrobacterium radiobacter* is another type of microorganism that protects young trees against infection by the crown gall bacterium.

Treat for insects and mites only when monitoring indicates that economically damaging populations are present or anticipated. Try to apply the pesticide when the target pest is at its most vulnerable state and when natural enemies are least affected. For example, treat the second rather than the first brood of the codling moth to avoid disrupting the natural enemies of aphids in the spring. When treating weeds after emergence, be sure they are at the growth stage recommended on the label.

Correct application of pesticides is most important for effective control. Calibrate the equipment properly, use the correct dosages and, where possible, place the chemical selectively for most efficient control. The recommended dosages and frequency of application given on pesticide labels may be higher than necessary for good control when properly timed and applied. Recommended dosages for herbicides usually have to be adjusted for soil type, climatic conditions, and irrigation method used. Consult your farm advisor for the lowest effective dosage and best application method for the pest problem in your orchard.

Before choosing a pesticide, check the latest University of California recommendations for walnuts. Always read the pesticide label; it outlines safety precautions, legal requirements, and registered uses for each product.

Problems Associated with Pesticides. The attributes that make pesticides valuable control tools may in some circumstances bring problems that directly influence the success of an IPM program. These problems include the development of pesticide resistance, pest resurgence, and secondary pest outbreaks. Pesticide applications may also affect neighboring crops, wildlife, and human health.

Pest Resurgence. Pest resurgence refers to the rapid rebuilding of a pest population that was recently controlled by a pesticide application (Figure 12). Such resurgence is especially a problem in pest populations that have short reproductive cycles, such as aphids and mites. Faced with resurging populations, you may have to spray repeatedly, increasing your production costs and the chances for creating secondary pest outbreaks.

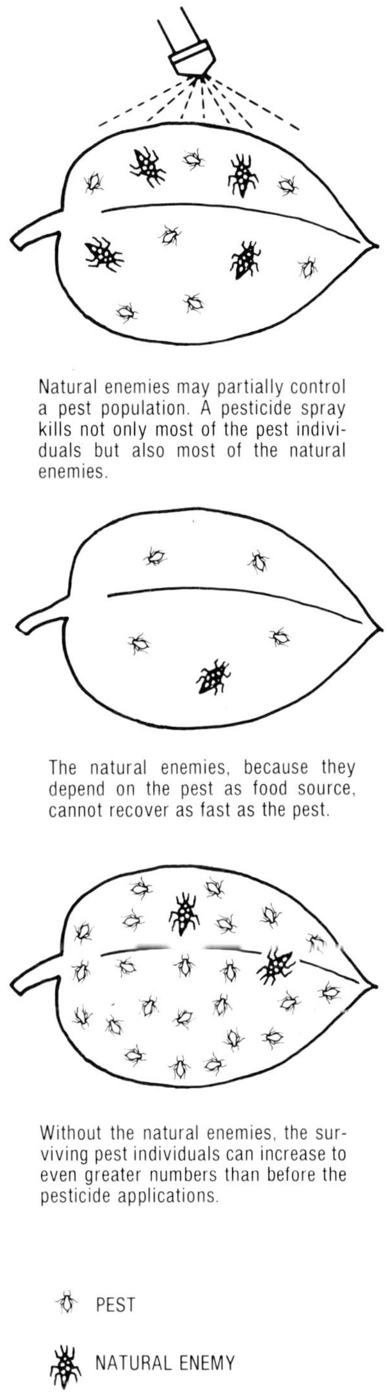

FIGURE 12. **Pest resurgence may occur when pesticide treatments for a target pest eliminate its natural control.**

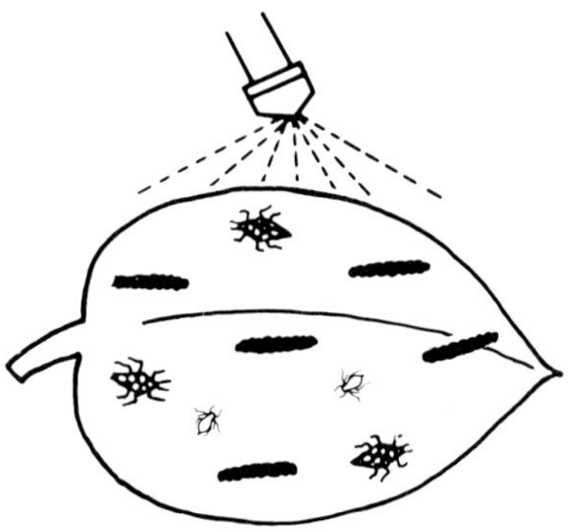

A pesticide is applied to control an economically damaging population of pest A. The spray controls pest A but also kills the natural enemies that are controlling pest B.

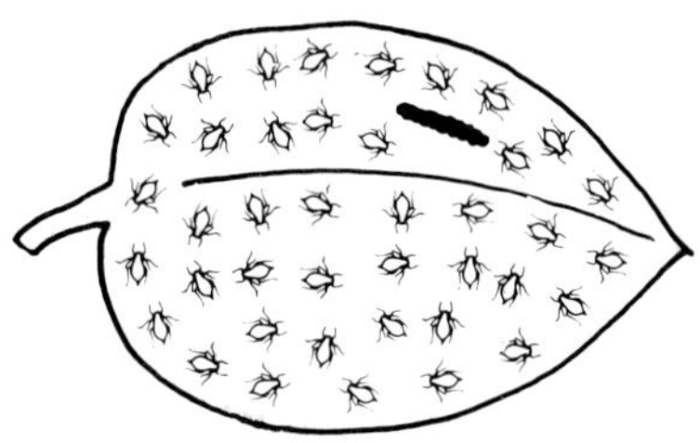

Released from the control exerted by natural enemies, pest B builds up to economically damaging levels.

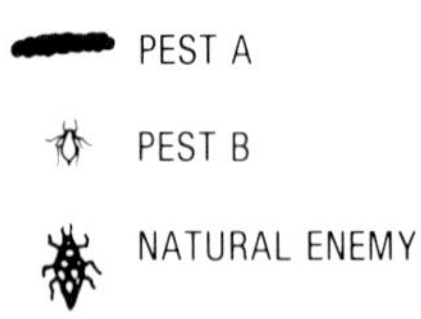

FIGURE 13. Secondary pest outbreaks among insect pests are most often induced by killing natural enemies. Disruption of competition, however, may also cause outbreaks of secondary weed, disease or insect pests.

Secondary Pest Outbreak. Secondary pest outbreak refers to the buildup of insects and mites that were previously controlled by natural enemies (Figure 13). In walnuts, treating for codling moth with certain materials in the early spring can cause outbreaks of the walnut aphid by killing the aphid's parasite but not the aphids. To reduce the chances for pest resurgence and secondary pest outbreak, choose selective materials when possible and time applications to be most destructive to the pest species but least destructive to the beneficial species.

Pesticide Resistance. Development of resistance to pesticides can occur in pest populations, particularly in those that reproduce rapidly, such as pathogens, aphids, and mites (Figure 14). Although less common, resistance has also been reported in some rodent and weed species. When resistance develops, it is desirable to alternate materials with a different mode of action when these are available. To slow the development of resistance, use pesticides only when needed and apply the materials selectively.

Hazards to Bees. Many insecticides used in walnuts are toxic to bees. Although bees do not pollinate walnuts, they may forage in the orchard on a blooming ground cover, on pollen or on honeydew excreted by aphids. If bees are foraging in the orchard, choose insecticides that are least toxic to bees and apply sprays in early morning or late evening when bees are not active. If you have to apply an insecticide that is toxic to bees, check with your agricultural commissioner; he will know if beekeepers in your area requested notification by collect telephone or telegraph. Inform these beekeepers of your scheduled spraying ahead of time. For details of toxicity categories of pesticides and spray precautions, see the UC Agricultural Science Publications 2883. Consult your agricultural commissioner for legal restrictions.

Hazards to People and Wildlife. Many of the pesticides used in walnuts are potentially hazardous to people and wildlife. Applicators are most heavily exposed to pesticides, and they may be at risk of acute poisoning or chronic ailments if adequate safety procedures are not followed. Minimize health problems by following label directions, wearing appropriate protective clothing, observing state and local regulations, including reentry intervals, and confirming the availability of emergency health care. To reduce drift, check the wind speed and direction and the calibration of the equipment. Avoid drift into ponds, rivers and other water supplies. The UC Agricultural Science Publication 2768 and booklet 4070 discuss health hazards and safety precautions in more detail.

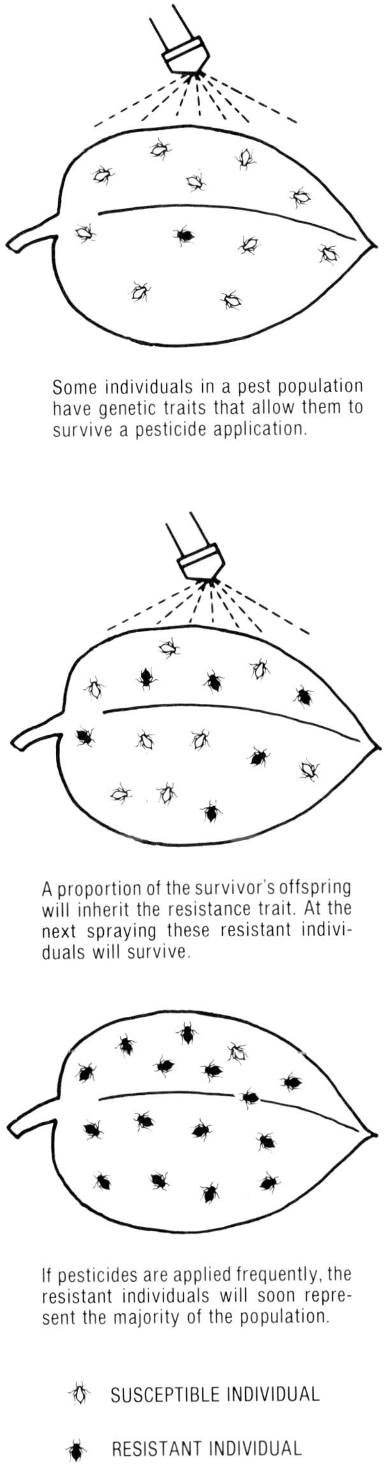

FIGURE 14. **Pest populations develop resistance to pesticides through genetic selection.**

Vertebrates

A few vertebrate species may move into walnut orchards for food and, to a lesser extent, for shelter. Ground squirrels are the most damaging rodent pest. They cause losses by feeding on the nuts and by digging extensive burrows in the orchards. The burrows may expose roots, disrupt the irrigation system and interfere with mechanical harvesting. In walnut orchards near weedy areas, tree squirrels may take large amounts of nuts. Statewide, they probably rank second to ground squirrels in causing damage to walnuts.

Other vertebrates are occasionally a problem in walnut orchards, depending on the growing district and local conditions. Gophers can present a problem when they dig burrows in the orchard though they generally do not feed on walnut trees. Meadow mice, rabbits and deer rarely cause damage because walnuts are less attractive to them than many other fruit or nut trees. Among the birds seen in walnut orchards, crows and occasionally magpies present a problem.

The potential for damage by vertebrates varies from orchard to orchard and depends largely on the region, the surrounding terrain and, to a lesser degree, on the management practices in the orchard. Orchards near rangeland, wooded areas or other uncultivated areas are more likely to be invaded or reinvaded by vertebrates than orchards bordering field crops. Flood irrigation discourages infestation by meadow mice, gophers and ground squirrels.

FIGURE 15. The California ground squirrel has a mottled brown to gray fur with flecks or a dark patch on its back.

Predators, diseases and available food sources all influence a vertebrate population. Predators, such as coyotes, foxes, badgers, snakes, hawks, and owls feed on many vertebrate species. These predators should be encouraged where they do not pose a hazard to farm animals. Natural control alone, however, cannot be relied on to prevent vertebrates from becoming agricultural pests at times.

A few preventive measures can make the orchard or the surrounding areas less favorable to pest invasion and survival. For example, removing debris, brush or prunings in or around the orchard will deprive ground squirrels of favored cover. Clearing weeds around tree trunks will discourage infestation by meadow mice. If the preventive approach proves unsatisfactory or unfeasible, other management methods, such as traps, toxic baits or fumigants, may have to be used.

Because each vertebrate pest must be managed differently, it is important to identify correctly the species that is causing the damage. Watch for the animals or indications of their presence, such as burrows, soil mounds, feces, nut shells or gnawing marks on the bark.

The timing of a control action is often critical. For instance, you can control ground squirrels most effectively after they have emerged from hibernation in late winter but before they have reproduced. At this time in early spring they do not accept cereal baits since they feed mainly on fresh vegetation. Therefore, fumigation is a better choice because the soil is moist and retains the toxic gas.

Because of the complexity of vertebrate pest problems, no numerical guidelines are available that specify what infestation levels and combination of environmental conditions justify control actions. Instead, to make management decisions for a vertebrate pest, the biological information presented in this chapter and the history of vertebrate pests in your orchard should be weighed against the cost and effectiveness of available control methods. After you have achieved control, establish a routine monitoring program to detect reinvasions and to make future control decisions easier and more efficient.

Ground Squirrels
Spermophilus beecheyi

Various subspecies of the California ground squirrel (*S. beecheyi*) are found in the walnut-growing areas of the state.

Ground squirrels feed on young nuts and mature nuts on the ground or in the tree. They can climb trees and strip branches of large numbers of nuts. Although they generally do not feed on other parts of the walnut tree, they can severely damage plastic irrigation lines by gnawing on them. Ground squirrel burrows in the orchard can disrupt irrigation and cause erosion. In some areas, ground squirrels present a potential health problem because they can carry human diseases, such as sylvatic (bubonic) plague, Rocky Mountain spotted fever, and tularemia.

The California ground squirrel weighs 0.5 to 0.7 kg (1 to 1.5 lbs.) as an adult. It is about the same size as a tree squirrel but has a much less bushy tail and a mottled fur with flecks or a dark patch on its back (Figure 15). Ground squirrels often bask in warmth of the morning or late afternoon sun. They are social animals, living in groups of up to 20 individuals.

When ground squirrels themselves are not seen, their burrows are a good indication of their presence. Each burrow system has several openings with scattered soil in front (Figure 16). In contrast, burrows of pocket gophers have crescent-shaped mounds and plugged entrances. Individual ground squirrel burrows may be 1.5 to 9 meters (5 to 30 feet) long, 75 cm to 1.2 meters (2.5 to

FIGURE 16. **Entrances to ground squirrel burrows are open and often have mounds of soil.**

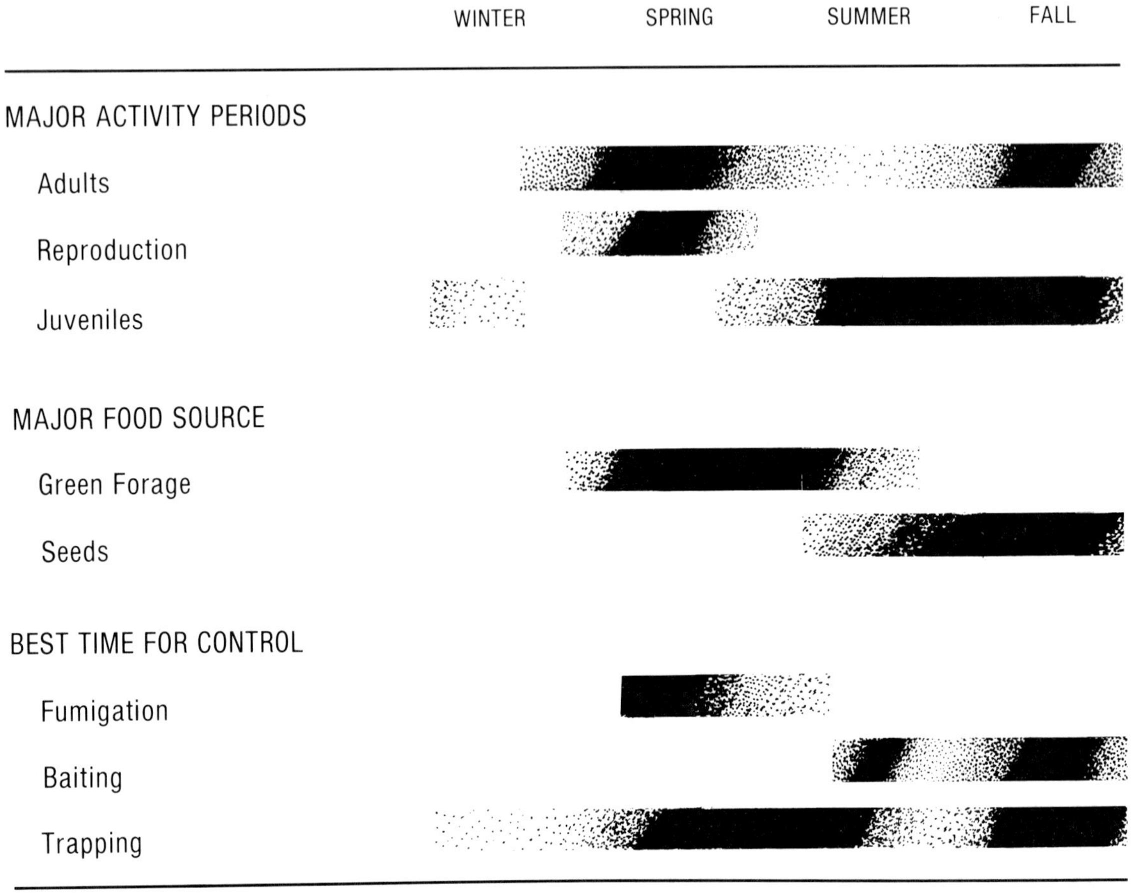

FIGURE 17. To choose the best time and type of control, consider the activity periods and food sources of the ground squirrel.

4 feet) below the surface, and about 10 cm (4 inches) wide. The burrows provide a place to retreat, sleep, hibernate, rear the young, and store food, including substantial quantities of walnuts.

Ground squirrels usually dig their burrows along ditches, fence rows, and on uncultivated land. Where orchards are not flood irrigated, the squirrels may also move into the orchard to dig their burrows.

Figure 17 shows the major activity periods and food sources of ground squirrels. They emerge from hibernation about late January in the southern part of the Central Valley and about a month later in the northern part. The females have one litter of six to eight young in the spring. About six weeks after birth, the young emerge to feed above ground. Ground squirrels prefer to feed on green vegetation in the spring and on seeds, grains, fruits and nuts during the summer and fall.

The adults often go into a temporary state of inactivity (aestivation) for part of the hot summer and into hibernation in the winter. The young usually do not aestivate or hibernate during the first year. Because of these periods of inactivity, squirrels seem to be more numerous in the spring and fall. Ground squirrels can live five years or more.

Management Guidelines

Whether and when you need to take management action depends largely on the local conditions. If your orchard is surrounded by infested, uncultivated lands, reinvasion will be a perennial problem. In cultivated areas where only a limited amount of highly desirable habitat exists, removing stumps, brush piles and debris can be helpful in limiting populations.

To monitor ground squirrels, check the perimeter of the orchard about once a month for animals or their burrows. Monitor during the early morning or late afternoon when squirrels are most active. Record the approximate number of squirrels and the location and number of burrows. You can expect a great increase in ground squirrel activities when the adults emerge from hibernation.

If your monitoring records indicate that the squirrel population is moving into the orchard, consider control methods, such as traps, fumigants and toxic bait. Choose the method carefully according to the time of year (Figure 17), area, and number of squirrels as well as the cost, labor and potential hazard of the treatment.

You can control ground squirrels most effectively in late winter or early spring after they have emerged from hibernation but before they have reproduced. In timing your control actions, keep in mind that males appear first; all animals have emerged from hibernation when the sex ratio is 1:1. At this time you may either use a fumigant or traps that do not require baiting. Baited traps or toxic bait provide good control later in the spring or in the fall when squirrels collect and eat seeds, grains and nuts. Attempts at controlling ground squirrels during the summer are not very successful because many of them are aestivating in closed burrow chambers. If you are in doubt about correct timing for bait acceptance, consult your agricultural commissioner or farm advisor.

Fumigants. Fumigants work best in the early spring when moist soil helps retain a high toxic gas level in the burrows. Several types of fumigants are available. Nonrestricted fumigants, such as gas cartridges (woodchuck or smoke bombs) are available from agricultural commissioners or private companies. When ignited, these cartridges release smoke and toxic gases, in particular, carbon monoxide. Following label instructions, place one or more cartridges in the active burrow opening and plug all other openings with soil. Insert and ignite the fuse, push the cartridge deep into the burrow, and quickly seal the entrance. Be careful not to inhale the smoke. Watch the area for escaping smoke and plug any leaks. Check the burrows after about three days. Treat again where opened burrows indicate that squirrels have dug out. Do not use cartridges where fire may be a hazard, such as near dry grass or beneath a building.

Restricted fumigants, such as carbon bisulfide and methyl bromide, require a permit from the agricultural commissioner. These materials can be very hazardous and should be applied only by experienced persons. They are toxic also to plants and can damage or kill walnuts if applied to burrows beneath trees. Apply these materials only to orchard margins away from trees and strictly follow all label instructions.

Traps. Trapping ground squirrels works well in small areas or where there are few squirrels. Be sure to place traps in such a way that pets and wildlife are unlikely to be caught.

One effective trap is the California 44 Pocket Gopher Trap, which can be modified to catch ground squirrels (Figure 18). Place a bait of walnuts, almonds, oats, barley or melon rind between the two traps under the

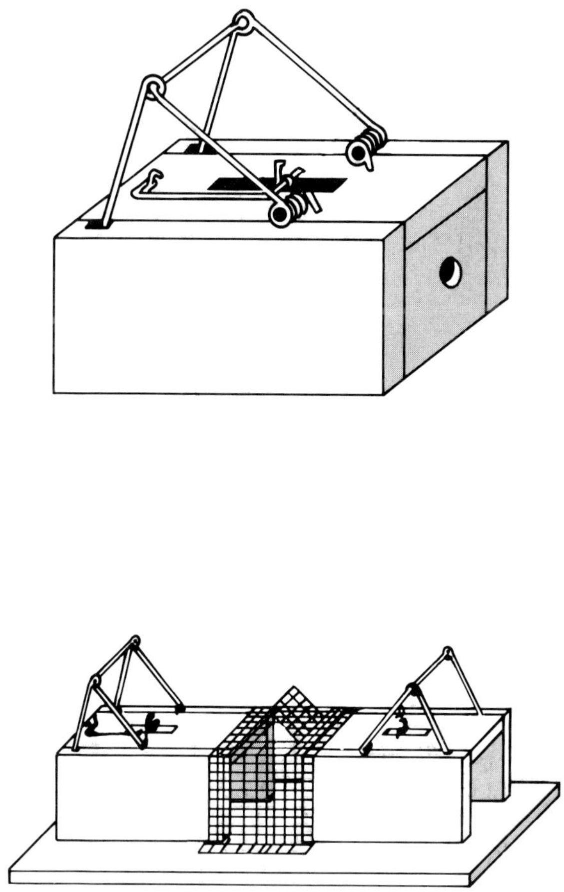

FIGURE 18. To modify California 44 pocket gopher traps for control of ground or tree squirrels, remove the back of the traps (top), connect two traps with wire mesh and attach them to a board (bottom).

FIGURE 19. You can make a bait station for ground squirrels out of wood, discarded tires or plastic pipe, as this one.

FIGURE 20. The eastern fox squirrel has a bushy tail and a reddish brown fur.

wire mesh to lure the squirrels through the trapping mechanism. Bait but do not set the trap for the first few days. After the squirrels have become accustomed to feeding at the station, set the trap.

Another trap, the Conibear®, does not require baiting; thus it is effective against ground squirrels all year. Live traps work well but the live animals have to be destroyed after they are caught.

Toxic Bait. Poison baits are the control method most commonly used against ground squirrels. Either single dose or multiple dose baits are effective, although each is used differently. Single dose poisons kill the animals after just one feeding and are usually more toxic. Depending on label instructions, the bait is generally distributed by hand over 2 to 3 square feet near burrow openings where the vegetation is most sparse. Do not pile bait; spreading it thinly takes advantage of the squirrels' natural foraging habits. Some baits are also formulated to be applied by vehicle-mounted seeders. After treating an area, removing and disposing of carcasses can prevent poisoning of dogs or other scavengers that may feed on them.

Multiple dose anticoagulant baits act more slowly, and in general the squirrels must feed on them for several days with no more than 48 hours between feedings. These baits cause fatal internal hemorrhaging, but they are considered less hazardous to human beings and nontarget species than some single dose baits because they produce delayed symptoms and antidotes are available.

Multiple dose baits are most often registered for use in bait stations. You can make bait stations from metal or plastic pipe, discarded car tires or wood (Figure 19). They should have entrance holes large enough to admit squirrels but not larger animals. Place the station near runways or burrows and secure them so that they cannot easily be tipped over. When you first set them out, inspect the stations daily and add bait as needed. Maintain the bait until feeding ceases. This process may take one to four weeks because many squirrels may not start feeding immediately and individual susceptibility to the toxicant varies. When you are finished, remove the unused bait from the stations.

Some anticoagulants are registered also for spot baiting or broadcasting. Lightly scatter bait near burrow openings or runways and reapply every other day for at least three applications, or as indicated on the label.

Occasional Vertebrate Pests

Tree Squirrels

Eastern fox squirrel *Sciurus niger,* Eastern gray squirrel. *Sciurus carolinensis.* In some walnut-growing areas, two introduced species of tree squirrels, the eastern fox squirrel and the eastern gray squirrel, occasionally cause economic damage by feeding on the nuts and collecting large quantities of nuts for storage. The eastern fox squirrel is much more common than the eastern gray squirrel and is well established in city parks, residential areas and adjacent agricultural lands, especially in Fresno, Sacramento and coastal areas south of San Francisco. The eastern gray squirrel presently occurs only in a few locations in the San Francisco Bay Area, the San Joaquin Valley, and a small portion of Calaveras County.

The two native tree squirrels are rarely a problem to walnut growers; the western gray squirrel (*Sciurus griseus*) seems to have been largely displaced by the eastern fox squirrel, and the Douglas tree squirrel (*Tamiasciurus douglasii*) lives in forests.

In contrast to ground squirrels, tree squirrels have a bushier tail and their fur is not flecked or mottled (Figure 20). Tree squirrels prefer to live in woody areas along creeks and river banks, from where they can invade adjacent orchards for food. Although they live mainly in trees, some tree squirrels, particularly the fox squirrel, spend considerable time foraging on the ground. The eastern gray squirrel feeds on a variety of green and ripe nuts, seeds, mushrooms and fruits. The eastern fox squirrel has a similar diet but also feeds on bird eggs and insects. Both species usually feed only in localized areas of orchards adjacent to wooded areas. The eastern fox squirrel can be distinguished from the eastern gray squirrel by its reddish brown color; the eastern gray squirrel is very similar in appearance to the western gray squirrel.

Tree squirrels nest in tree holes or build nests of sticks and leaves in the branches of trees. The females of the eastern gray squirrel bear two litters per year of three to five young, one in late winter and one in late summer. Yearling females of the eastern fox squirrel bear one litter; older females have two litters per year between January and April and July and September.

Management Guidelines. To control the eastern gray squirrel, you need a permit from your local game warden or the regional office of the Department of Fish and Game. Poisoning of this species is illegal in California. When the fox squirrel causes damage, it can be taken in any manner; no permit is required.

Trapping is often sufficient to control tree squirrels. The California 44 Pocket Gopher Trap, modified as for control of ground squirrels, can also be used for tree squirrels. Fasten one or two box traps on a baseboard and nail it on top of a horizontal limb in a tree where you detect feeding damage (Figure 21). Place a handful of nutmeats well behind the trigger mechanism to attract the squirrels. You can remove a considerable number of tree squirrels with relatively few traps if you keep them in continuous operation while squirrels are feeding on walnuts.

FIGURE 21. For control of tree squirrels, place modified gopher traps on limbs where damage occurs.

Tree squirrels can also be controlled by shooting; consult your local game official for regulations concerning the shooting or hunting of squirrels.

Pocket Gophers *Thomomys* spp.

Pocket gophers occasionally invade walnut orchards, especially those with a ground cover or sprinkler or drip irrigation; flood irrigation tends to discourage the buildup of a gopher population. Gophers feed mostly on the roots and stems of field crops and weeds; they rarely damage walnut trees. They are a concern to walnut growers mainly because they dig burrows in the orchard. The burrow mounds interfere with mowing or harvesting operations and the tunnels divert irrigation water.

Pocket gophers are rarely seen, but characteristic crescent-shaped mounds of soil indicate their presence. The gophers dig tunnels 10 to 30 cm (4 to 12 inches) below the surface, push the soil out lateral exits and plug these openings. On irrigated land, gophers breed almost throughout the year, and the female may bear three litters averaging 5 to 6 young.

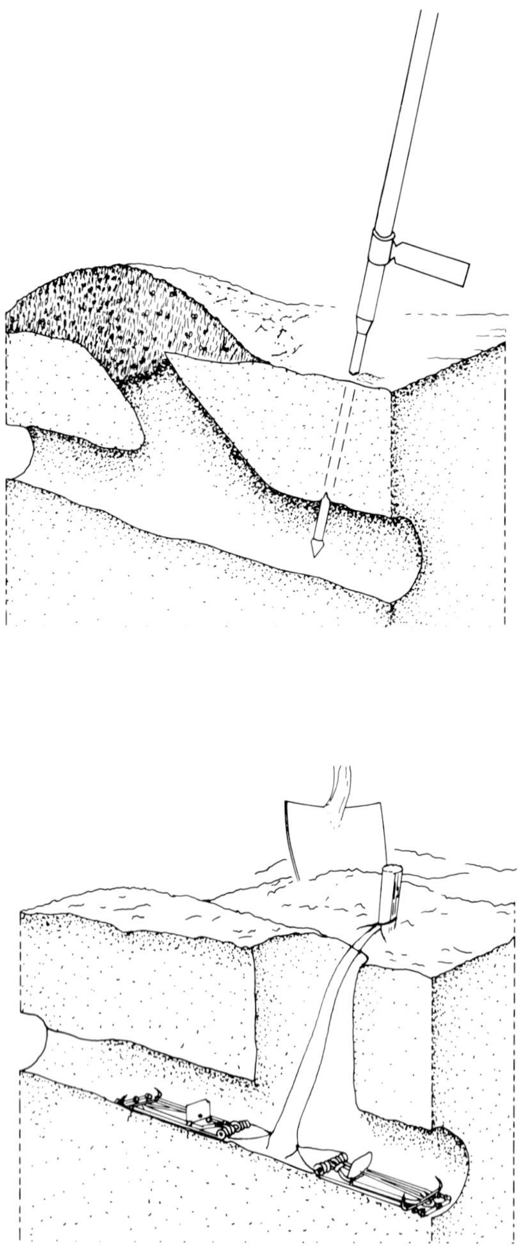

Figure 22. Control of pocket gophers using traps. Locate the main burrow with a metal probe (top); place two traps in opposite directions, fasten them to a stake and cover the hole (bottom).

Management Guidelines. Monitor gophers in late fall, winter or spring. As the ground becomes softened by rain, the gophers increase their digging activity, and you can easily detect fresh mounds. In the spring, you may find new gopher mounds in various parts of the orchard as the young set up burrow systems of their own. Check the orchard about once a month, especially the perimeter where gophers may move in from adjacent fields or orchards.

If your monitoring records indicate that a gopher population is already in the orchard or is expanding, you have two control options: traps or toxic bait. Fumigants do not work well because the burrow system is extensive and gophers can quickly seal their tunnel when they detect the poisonous gas.

Traps are useful for a few gophers in a small area and are most effective if they are inserted into the main tunnel between two fresh mounds. Locate the main tunnel, which usually extends from the lower side of the mound, with a steel rod 6 mm (¼ inch) wide or a specially designed metal probe. Place traps as shown in Figure 22. For most effective control, inspect the traps twice daily.

You can also control pocket gophers with bait treated with a rodenticide. Check with your agricultural commissioner or farm advisor for availability of baits. To place bait by hand, locate the main burrow, enlarge the probing hole with the metal rod and drop the bait into the tunnel, following label instructions. Cover the hole to prevent soil from dropping onto the bait. Repeat once or twice along the runway. Handle poison bait with extreme care to avoid spilling bait on the surface, where it may cause accidental poisoning of nontarget animals. Keep bait out of reach of pets and children.

When a large population of gophers needs to be treated over a wide area, mechanical baiting will give the most effective control. For details see the U.C. Agricultural Science Publication 2699.

Birds

A number of bird species may be attracted to the walnut orchard for food, nesting and cover. Most of these are not considered pests. However, crows and occasionally magpies may take large amounts of walnuts in localized areas. Scrub jays and sapsuckers may frequent orchards but rarely cause economic damage.

Crows join into flocks for most of the year and establish a central roost, from which they fly each day to feeding areas. Crows start feeding on half-mature nuts and continue through harvest. During the winter, they feed on mummy nuts left in the orchard. Aside from walnuts, crows feed on seeds, fruits, other nuts and also on insects and even on mice. Magpies feed in small groups on insects, small vertebrates and eggs and supplement their diet with fruits, grains and nuts.

A variety of frightening devices, such as carbide exploders, shell crackers or distress calls, can be successful in keeping the birds temporarily away from the orchard. Shooting crows is another control option, but it requires some skill because crows are intelligent and wary birds. Toxicants have been used to control crows. Trapping is not practical. Crows are classified as migratory birds in the Code of Federal Regulations; however, to control crows, you do not need a permit if they damage or are about to damage your crop. For details about regulations and control options, consult your agricultural commissioner.

Other Vertebrates in Walnuts

Meadow mice (*Microtus* spp.), also called meadow voles or field mice, inhabit roadsides, canal banks and fencerows. They may move into walnut orchards and feed on young tree bark when weeds around the tree trunks offer protection. However, meadow mice generally avoid the rootstocks commonly used for walnuts, and mouse damage in walnuts is rare compared to other deciduous fruit and nut orchards.

To make a young orchard and its surroundings less favorable to mice infestation, mow the ground cover or cultivate frequently and control weeds around tree trunks and along fencerows, roadsides and ditchbanks.

Among the hares, the black-tailed jackrabbit (*Lepus californicus*) is most common as an agricultural pest. The true rabbits, the cottontail and brush rabbit (*Sylvilagus* spp.) can, on very rare occasions, be a pest where trees are newly planted. Hares and rabbits may feed on the bark of young walnut trees, usually only during the winter when other, more preferred, foods are limited. As a rule, hares and rabbits are of little concern to walnut growers.

Deer occasionally damage newly planted walnut orchards that are located near natural habitats, such as wooded areas and stream side thickets. They feed at night on young tree foliage and may break small limbs. Walnut trees, however, are less desirable food for deer than other fruit and nut trees.

Insects and Mites

Insect and mite pests in walnuts can be grouped as key, secondary and occasional pests according to their potential for causing economic damage and their frequency of occurrence. The number of pests in each of these groups and their damage potential differ regionally and, to a large extent, according to the cultivars planted.

Key pests are those that have the greatest potential for economic impact on the tree or the crop. They usually damage nuts directly and are not well controlled by natural enemies. They are present in most walnut-growing areas most of the time. The key pests in California walnuts are the codling moth, the navel orangeworm, and the walnut husk fly.

In the spring, the first generation of the codling moth larvae infests nutlets, causing them to drop from the tree; the second and sometimes the third generations cause damage in the summer by feeding on the kernels. The navel orangeworm builds its populations in the spring and summer on damaged nuts. Like the codling moth, the navel orangeworm feeds on kernels, but this pest can enter sound nuts of the new crop only after the husk has split. The walnut husk fly is a pest in summer and early fall. Its maggots feed only on the husk, but the feeding turns the husk black, staining the shell and causing a grade reduction for inshell sales. An early infestation can also lower the nut quality.

Management of the codling moth and the husk fly depends on chemical treatment, determined by a careful monitoring program. Managing the navel orangeworm depends largely on preventive measures, such as destroying mummy nuts and trash nuts, controlling codling moth and walnut blight, and harvesting early.

Secondary pests can inflict economic damage in some orchards in some years but are not annual problems in most orchards. These pests, including aphids, scale insects and plant-feeding mites, damage leaves and lower tree vigor, thereby reducing kernel quality and yield. In some cases, outbreaks of secondary pests occur when insecticides applied to control key pests kill the natural enemies of aphids, scale insects and mites.

Two aphids, the walnut aphid and the dusky-veined aphid, are secondary pests in walnuts. Both feed on leaves; their populations can cause premature leaf drop and a reduction in nut size, quality and yield. The walnut aphid occurs throughout the state and was an important pest until the early 1970s when a parasitic wasp was successfully established for its control. The dusky-veined aphid was once restricted to the coastal areas, but populations have recently increased in the Sacramento and the San Joaquin Valleys. Populations in these areas were probably suppressed by competition with the walnut aphid and the chemicals applied for walnut aphid control.

Three soft scales, the frosted scale, the European fruit lecanium, and the calico scale, can appear in heavy populations when their natural control has been disrupted by insecticides or adverse weather. They feed on leaves or twigs throughout the season. They are all similar in biology and feeding habits, but the frosted scale is found most frequently.

The armored scales that infest walnuts feed mainly on bark. The most common is the walnut scale, but the San Jose scale can be a problem in localized areas of an orchard. The oystershell and Italian pear scales are less frequent problems. Heavy infestations of any of these scales reduce tree vigor and sometimes kill limbs.

Plant-feeding mites are widespread, but the species present on walnuts and their relative importance vary regionally. In the Central Valley, the two-spotted mite and the Pacific mite are important. The European red mite is often found in the cooler coastal regions and in shady orchards in the Sacramento Valley. Outbreaks are usually induced by insecticides, and certain orchard conditions tend to promote mite problems. Dust from roadsides, frequent cultivation or flailing often contribute to mite outbreaks. A diverse complex of predators feeds on the various mites and in many cases keeps them below damaging levels if not disrupted by chemical treatment for other pests.

A number of walnut pests cause only occasional problems. The redhumped caterpillar, fall webworm, and false chinch bug feed on leaves and can defoliate walnut trees, especially young ones. The redhumped caterpillar and the fall webworm prefer the warmer interior growing regions. The earwig and the filbertworm may be found in walnut orchards but they rarely reach damaging levels. The flatheaded borer occasionally bores into young trees, but usually only into wood damaged by sunburn or diseases. The fire ant may infest walnuts after they have been shaken from the tree. Rapid removal from the ground can avoid such infestation.

Monitoring Insects and Mites

A good monitoring program for insects is essential to the implementation of integrated pest management. Use your monitoring results together with field history to determine the need for management actions. It is important to identify pests and their associated damage correctly because even closely related species may require different control measures. The photos in this manual will assist in identifying the most common insects and mites found in walnut orchards. For additional help, consult your farm advisor, agricultural commissioner's staff or PCA.

You need to carry out precise sampling programs for individual species only at certain times of the year. Figure 23 shows the best monitoring times for major insect and mite pests in walnuts. Begin monitoring well before potential problems normally occur and continue through the pest's damaging stages. Some pests are best monitored during the dormant period. Check every orchard at least once a week when pests are likely to occur and more often when populations are expected to rise to damaging levels. Keep written records of all monitoring results. Figures 28 and 33 show examples of sampling forms for codling moth and aphids.

A monitoring scheme should include all sections of the orchard. The sampling unit depends on the size of the orchard; divide large orchards into 20-acre sampling blocks with about five sampling sites chosen at random.

At these sites, monitor aphids on leaves and mites and scale insects on leaves or twigs. Be aware that mites may start in "hot spots," particularly around the edges of an orchard. Sample these areas independently from the rest of the orchard.

Monitoring programs for the codling moth and the husk fly are based on trapping. The number of traps needed depends on the size of the orchard. For convenience, place the traps for the two species in the same or nearby trees at a sampling site.

Several other factors are important in monitoring insects and mites. Different species or different life stages prefer certain parts of the tree or the orchard. For example, walnut aphids feed on the lower surface of leaves, dusky-veined aphids on the upper surface. Husk flies prefer vigorously growing trees with dense foliage. The first instar nymphs of soft scales feed on the lower side of leaves during the summer; second instar nymphs and mature scales feed on twigs and branches. Conditions favorable for pest development vary between neighboring orchards due to different cultural practices or due to the age and cultivar of the orchard. Keep separate records for each orchard, noting special conditions and pest problems.

It is important to monitor for natural enemies as well as for pest species. High populations of natural enemies can rapidly bring a potentially damaging insect or mite population under control. Look for predators or note the percentage of parasitized individuals in the pest population.

Monitoring Methods

Traps. Traps attract insects by scent, color, shape or ultraviolet light. They can help you detect activity peaks or check for the presence of adult insects in the orchard. You can use trapping information to time pest control actions more effectively.

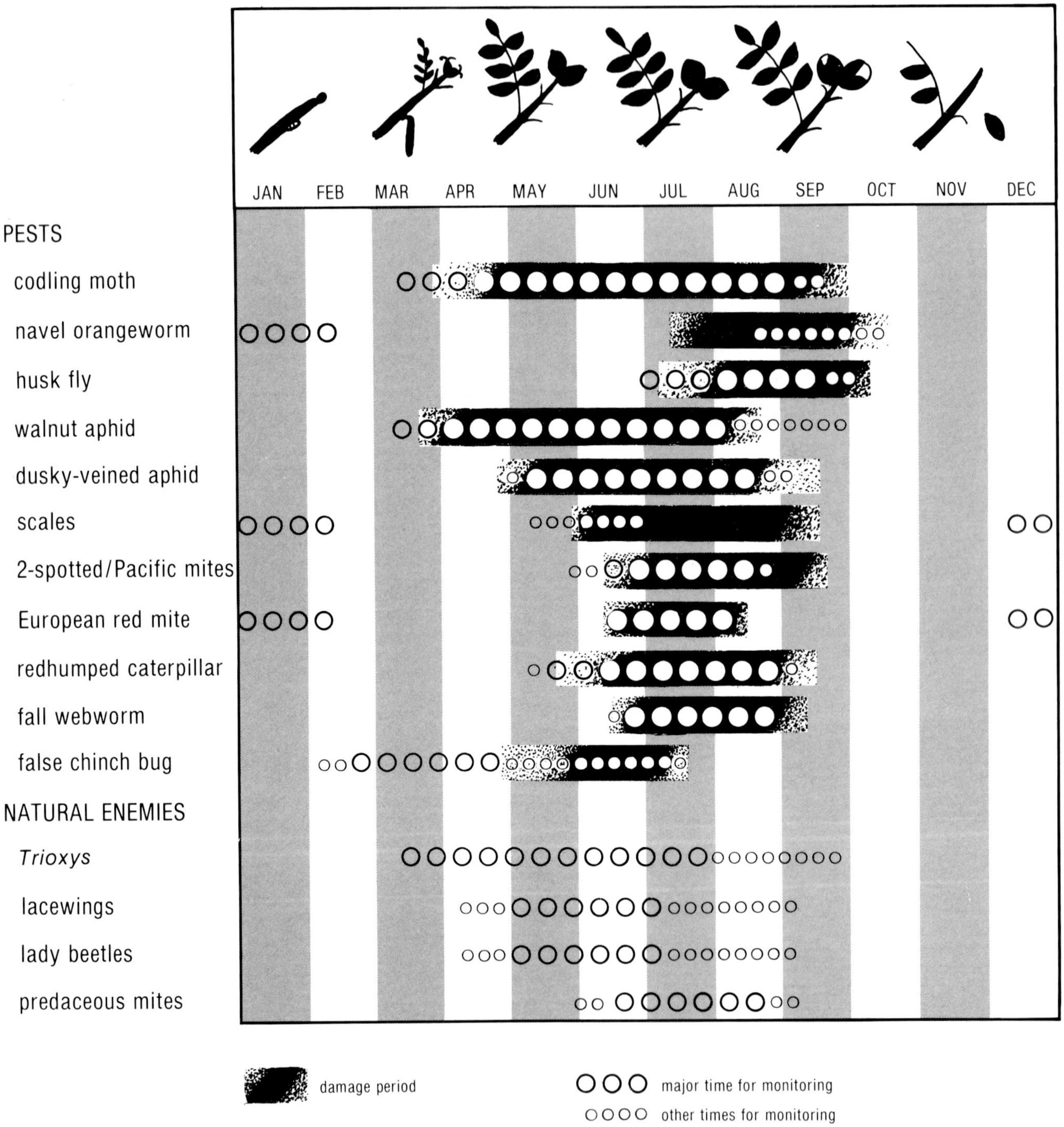

FIGURE 23. Monitor insect and mite pests when they are likely to cause damage and/or at other times as indicated. Also check for natural enemies. Developmental periods are for an early cultivar; events may occur up to 1 month later in late cultivars.

FIGURE 24. Use a pheromone trap to monitor the flight activity of the codling moth.

A pheromone trap is used to monitor the activity of the codling moth (Figure 24). Programs using pheromone traps are being tested for the San Jose scale. Pheromones are attractants produced by insects to communicate with individuals of their own species. Several sex pheromones, which may be released by the male or female to attract the opposite sex, have been identified and synthesized for use in pheromone traps.

Most commercially available pheromone traps are made of folded cardboard cards and are coated on the lower inside with a sticky substance and a replaceable plastic or rubber cap impregnated with a pheromone. The scent is slowly released from the cap and attracted insects get stuck in the trap. Sex lures are fairly specific although other species are sometimes caught. Pheromone traps require regular changing of the caps and the sticky liners, but they are easier to service than most other trap designs.

Some problems can arise in evaluating pheromone trap catches of the codling moth. Because only males are caught, counts may not reflect important female activities, such as egg laying. Guidelines in this book will help you adjust for this discrepancy. Flight activity varies with environmental conditions; trap catches should therefore be correlated with weather information.

There are various bait traps to monitor the husk fly, but the best trap available is a yellow rectangular cardboard coated with a bait and a sticky substance (Figure 25).

FIGURE 25. Monitor the husk fly with a yellow cardboard card impregnated with a bait and a sticky substance.

FIGURE 26. When using a hand lens, hold the lens close to your eye and move the object being viewed close until it is in focus.

Visual Counting. You can assess populations of aphids, mites and their natural enemies by counting the number of individuals present on a predetermined number of leaflets. Check for mature scale insects and their natural enemies by selecting an area of the trunk, limbs or branches and observing the number of scales and the percentage of parasitized ones. Redhumped caterpillar, fall webworm, false chinch bug and minor insect pests are also monitored visually. These methods are described in more detail in the individual pest sections. Figure 26 shows the correct use of a hand lens.

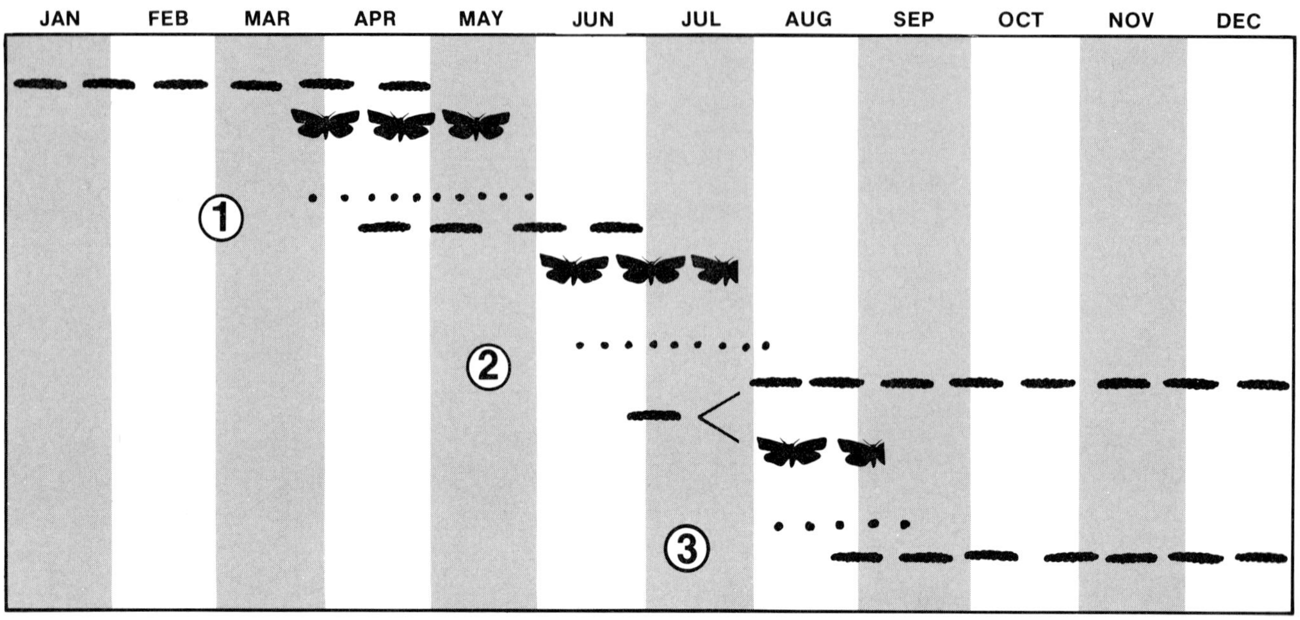

FIGURE 27. Seasonal development of the codling moth. Moth flights and developmental times vary with temperature and location. (1) First generation, (2) second generation, and (3) third (overwintering) generation. A portion of the second generation larvae enter diapause and become part of the overwintering larval population. Adults emerging from March through May are the first flight of moths.

Codling Moth
Laspeyresia pomonella

The codling moth is a major insect pest of apples, pears and walnuts. A distinct host race has evolved that prefers walnuts and has a life cycle well synchronized with the development of early walnut cultivars. It causes economic losses mainly in orchards planted with early season cultivars, such as Payne, Ashley and Chico. Nuts of these cultivars mature early enough to support the first generation larvae.

Controlling codling moth is the key to managing other insect and mite pests in walnuts. No adequate cultural or biological controls are currently available, so management depends entirely on the use of insecticides. Unfortunately, codling moth insecticides frequently kill the natural enemies of other pests, such as aphids, mites and scales, making it necessary to treat for these secondary pests. Intensive research, however, has led to a better understanding of the life cycle of the moth and to a monitoring program that makes it possible to decide whether and when treatment is necessary. This program provides more effective control of the codling moth and often reduces insecticide applications and the chance of secondary outbreaks.

Description and Seasonal Development

Codling moths overwinter as full-grown diapausing larvae in thick, silken cocoons under loose scales of bark or in trash on the ground near the trunk. The larvae need a period of cold temperatures before they change inside their cocoons to brownish pupae in early spring. Moth emergence usually coincides with the leafing-out of early walnut cultivars, which occurs from mid-March to early April in the Central Valley and from late March to mid-April in the coastal counties. During the day, the moths rest on branches and trunks. Their gray mottled appearance blends well with that of the bark, making them very difficult to detect. Codling moths can be distinguished from other moths likely to occur in the orchard by the coppery band at the tips of their wings.

Moth activity and growth require a certain threshold temperature, and the completion of one generation from egg to egg requires the accumulation of at least 1000 degree-days. Adults of each generation tend to be active at about the same time. These activity periods are called flights. Offspring of adults from the same flight generally occur in broods. In coastal and Delta areas where temperatures can fluctuate widely, flights and broods often are not as distinct as they are in the inland valleys.

Each overwintered female deposits about 60 eggs singly on leaves near nuts or on nuts themselves. They

ORCHARD LOCATION ______________________ BOTTOMS CHANGED ______________________

TRAPPING PERIOD ____________ to ____________ CAPS CHANGED ______________________

COMMENTS ____

	INDIVIDUAL TRAP COUNTS (moths/trap)					AVERAGE TRAP COUNTS		
DATE	trap 1	trap 2	trap 3	trap 4	trap 5	moths/trap	moths/ trap/day	accumulated moths/trap

FIGURE 28. Check your pheromone traps every three to four days and note the moth counts on a record sheet, such as this one. The average number of moths/trap/day and the accumulated moths/trap are important for making treatment decisions.

are disc-shaped and opaque white when first laid. They later develop a red ring and just before hatching the black head of the larva becomes visible. Eggs of the overwintered generation hatch after 5 to 20 days, depending on the temperature, and young larvae bore through the blossom end into nutlets. On Payne-type cultivars, larval infestation begins when the nutlets average about 1 cm (3/8 to 1/2 inch) in diameter. The larva of the codling moth is similar in appearance to that of the navel orangeworm but has no brown, crescent-shaped marks on the second segment behind the head. The larvae feed and complete their development inside the nut and pupate under loose bark or on the ground. Figure 27 shows a typical seasonal development of the codling moth. Adults of the first generation begin to emerge from the end of May to as late as the last week of June in the Central Valley, depending on the season. In the coastal areas, emergence begins in late June to early July. These adults lay the eggs of the second generation. Because of the higher temperatures, the eggs and larvae of this generation develop faster than those of the overwintered generation. Newly hatched larvae of the second generation bore into husks anywhere on their surfaces. After maturing, a proportion of these larvae leaves the nuts and enters an inactive state called diapause until the following spring. The remaining larvae develop into adults, which begin to emerge by late July or the beginning of August. They produce a partial third generation, which can cause significant damage in some seasons. When they are fully grown, these larvae leave the nuts and move to tree trunks or debris to spin cocoons and overwinter. Occasionally some may remain in the nuts if they are harvested before the larvae have matured. However, most larvae found in nuts at harvest are navel orangeworms.

Damage

The damage caused by the codling moth is different with each generation. First generation larvae reduce yield directly by causing nutlets to drop from the tree. They also serve as a breeding site for the navel orangeworm.

The adult of the codling moth can be distinguished from other moths in the orchard by the coppery band at the tip of the wings.

The egg is disc-shaped and opaque white when first laid; just before hatching, the black head of the young larva becomes visible.

The larva of the codling moth does not have a pair of crescent-shaped marks on the second segment behind the head; this distinguishes it from the larva of the navel orangeworm.

First generation larvae bore through the blossom end and feed on the developing nut.

Damaged nutlets have frass at the blossom end. Be careful not to confuse nuts damaged by codling moths with unpollinated nutlets or with blight-infected nutlets, which have dark lesions but no frass. Damage is generally most severe on early season cultivars.

Nuts attacked by the second and third generations remain on the trees but are unmarketable because of the feeding damage to the kernel. These damaged nuts can also serve as a breeding site for the navel orangeworm. You can often detect codling moth infestations by looking for frass produced by the larvae at the point of entry into the husk. Second generation larvae often enter through the side of the husk where the two nuts touch. After the shell hardens, the larvae enter the nuts through the soft tissue at the stem end.

Management Guidelines

Prevention. Aside from the use of insecticides as described below, there is little you can do to prevent damage by the codling moth in your orchard. Late season cultivars are less likely to be attacked. No effective biological control agents or cultural practices are available.

Monitoring and Control. The codling moth IPM program focuses on monitoring the moths to determine if populations are high enough to require treatment and to time applications so they will kill the first hatching larvae of the brood. Use pheromone traps (Pherocon® 1C) to monitor moth flights (Figure 24). The trapping density and the threshold levels presented here were adapted from the Culver method. This method was developed for the San Joaquin Valley and is currently being validated in other walnut-growing areas.

The number of pheromone traps needed for an orchard depends on its size; use at least 2 traps for an orchard of up to 10 acres; for 11 to 100 acres, use 1 trap per 10 acres. Larger orchards require 1 trap per 20 acres. Place the traps in trees away from the trunk, 6 feet above the ground and at least 100 feet into the orchard. At the beginning of each flight, always use fresh pheromone caps and bottoms. The caps and bottoms should be changed after about 4 weeks; change bottoms sooner if they become dirty or a total of 200 moths has been caught. Also, stir the sticky substance to maintain the adhesive quality. Record the moth counts as outlined in Figure 28. A graph is helpful to visualize the flight pattern throughout the season (Figure 29).

To forecast flights, accumulate degree-days. Record the daily minimum and maximum temperatures and look up these values for each day in Figure 30; the point where the two columns of the minimum and maximum temperature intersect in the middle of the table is the number of degree-days accumulated for that day. Sum the degree-days for successive days as shown in Figure 31.

The mature larva overwinters and pupates in a silken cocoon under loose bark; this cocoon was opened to show the pupa.

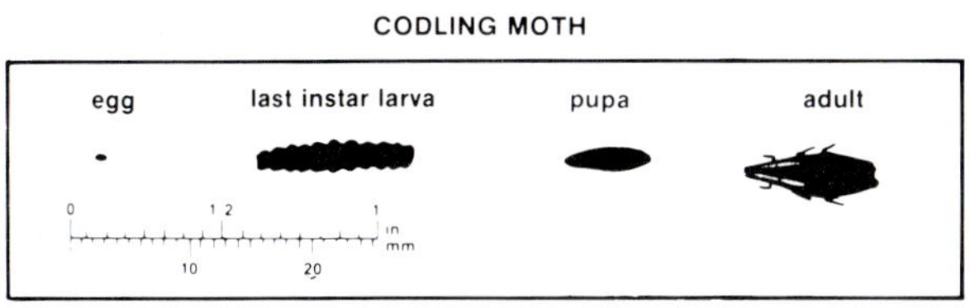

First brood larvae enter nuts through the blossom end. Feeding larvae produce frass, which protrudes from the entry hole (right nut) and distinguishes codling moth damage from blight damage (left nut).

Males are attracted to the pheromone trap and get stuck on the sticky bottom. Occasionally other moth species are attracted by the pheromone; here four *Grapholitha* moths were caught together with a codling moth (larger one on top).

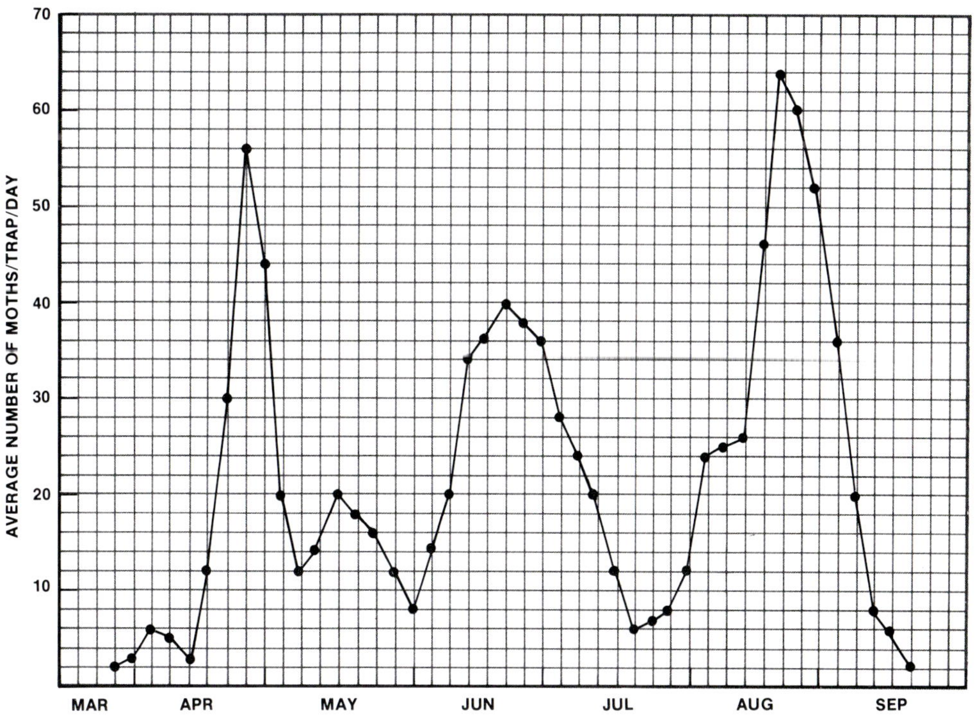

FIGURE 29. Keep track of the codling moth flights in your orchard on a graph such as this one. Each time you check your traps, plot the average moths/trap/day (Figure 28) on the graph. The tallest peaks indicate the peaks of the first, second and third flight. Use degree-day accumulations to determine the end of each flight.

DAY DEGREES ABOVE 50.0 F. AND BELOW 90.0 F.

MAX ↓ / MIN -->	26	28	30	32	34	36	38	40	42	44	46	48	50	52	54	56	58	60	62	64	66	68	70	72	74
118	21	21	21	22	22	22	23	23	24	24	25	25	26	27	28	29	29	30	31	32	33	33	34	35	36
116	20	21	21	21	22	22	22	23	23	24	24	25	26	27	28	28	29	30	31	32	33	33	34	35	36
114	20	20	21	21	21	22	22	23	23	24	24	25	25	26	27	28	29	30	31	32	32	33	34	35	36
112	20	20	20	21	21	21	22	22	23	23	24	24	25	26	27	28	29	30	30	31	32	33	34	35	35
110	19	20	20	20	21	21	22	22	22	23	24	24	25	26	27	28	29	29	30	31	32	33	34	34	35
108	19	19	20	20	20	21	21	22	22	23	23	24	25	26	26	27	28	29	30	31	32	33	33	34	35
106	19	19	19	20	20	20	21	21	22	22	23	23	24	25	26	27	28	29	30	31	32	32	33	34	35
104	18	19	19	19	20	20	20	21	21	22	22	23	24	25	26	27	28	29	29	30	31	32	33	34	35
102	18	18	18	19	19	20	20	20	21	21	22	23	23	24	25	26	27	28	29	30	31	32	33	34	34
100	17	18	18	18	19	19	20	20	20	21	22	22	23	24	25	26	27	28	29	30	31	32	32	33	34
98	17	17	18	18	18	19	19	20	20	21	21	22	23	24	25	25	26	27	28	29	30	31	32	33	34
96	16	17	17	17	18	18	19	19	19	20	21	21	22	23	24	25	26	27	28	29	30	31	32	33	34
94	16	16	16	17	17	18	18	18	19	19	20	21	21	22	23	24	25	26	27	28	29	30	31	32	33
92	15	16	16	16	17	17	17	18	18	19	19	20	21	22	23	24	25	26	27	28	29	30	31	32	33
90	15	15	15	15	16	16	17	17	17	18	19	19	20	21	22	23	24	25	26	27	28	29	30	31	32
88	14	14	14	14	15	15	16	16	16	17	18	18	19	20	21	22	23	24	25	26	27	28	29	30	31
86	13	13	13	14	14	14	15	15	15	16	17	17	18	19	20	21	22	23	24	25	26	27	28	29	30
84	12	12	12	13	13	13	14	14	15	15	16	16	17	18	19	20	21	22	23	24	25	26	27	28	29
82	11	11	11	12	12	12	13	13	14	14	15	15	16	17	18	19	20	21	22	23	24	25	26	27	28
80	10	10	11	11	11	11	12	12	13	13	14	14	15	16	17	18	19	20	21	22	23	24	25	26	27
78	9	10	10	10	10	11	11	11	12	12	13	13	14	15	16	17	18	19	20	21	22	23	24	25	26
76	8	9	9	9	9	10	10	10	11	11	12	12	13	14	15	16	17	18	19	20	21	22	23	24	25
74	8	8	8	8	8	9	9	9	10	10	11	11	12	13	14	15	16	17	18	19	20	21	22	23	24
72	7	7	7	7	8	8	8	8	9	9	10	10	11	12	13	14	15	16	17	18	19	20	21	22	
70	6	6	6	7	7	7	7	8	8	8	9	9	10	11	12	13	14	15	16	17	18	19	20		
68	5	5	6	6	6	6	6	7	7	7	8	8	9	10	11	12	13	14	15	16	17	18			
66	4	5	5	5	5	5	6	6	6	6	7	7	8	9	10	11	12	13	14	15	16				
64	4	4	4	4	4	4	5	5	5	5	6	6	7	8	9	10	11	12	13	14					
62	3	3	3	3	4	4	4	4	4	5	5	5	6	7	8	9	10	11	12						
60	2	2	3	3	3	3	3	3	3	4	4	4	5	6	7	8	9	10							
58	2	2	2	2	2	2	2	2	3	3	3	3	4	5	6	7	8								
56	1	1	1	1	1	1	2	2	2	2	2	2	3	4	5	6									
54	1	1	1	1	1	1	1	1	1	1	1	2	2	3	4										
52	0	0	0	0	0	0	0	0	0	0	1	1	1	2											
50	0	0	0	0	0	0	0	0	0	0	0	0	0												
48	0	0	0	0	0	0	0	0	0	0	0	0													
46	0	0	0	0	0	0	0	0	0	0	0														
44	0	0	0	0	0	0	0	0	0	0															
42	0	0	0	0	0	0	0	0	0																
40	0	0	0	0	0	0	0	0																	
38	0	0	0	0	0	0	0																		

FIGURE 30. Degree-days for the codling moth. To find the number of degree-days accumulated for a day, look up the minimum and maximum temperature for that day in the margins; the intersection of the two columns in the middle of the table gives you the degree-day total for that day.

The monitoring and management program for the codling moth is outlined below:

Moths of Overwintered Generation (First Flight)

- install an indicator trap mid-March; check every 3 days
- when the first moth is caught, install remaining traps and record daily minimum temperatures; start accumulating degree-days from the point when trap catches have begun to increase consistently
- record trap catches and remove moths every 3 to 4 days; stir adhesive regularly
- when peak flight approaches (at 200 °D or 18 to 25 days after start of consistent flight), consider accumulated trap catches; if the total number of moths/trap caught during this flight period is
  - — 50 or less, no treatment is necessary at this time except where navel orangeworm is a problem; treat at 225 °D or when Payne-type nuts average about 1 cm (3/8 to 1/2 inch) in diameter
  - — more than 50, treatment is necessary, but do not treat now unless navel orangeworm is a problem; delay until second flight to preserve the walnut aphid parasite
- continue trap counts every 3 to 4 days until 4 weeks after the start of the first flight
- continue degree-day accumulation until the second flight begins

Moths of First Generation (Second Flight)

- at about 950 °D after the start of the first flight (early to mid-June), replace bottoms and caps
- record trap catches and clean traps every 3 to 4 days
- once trap catches are increasing consistently, start accumulating degree-days from zero again
- if at the first or second check after a consistent increase in trap catches, the average number of moths/trap/day is 10 or more, treat as soon as possible (no later than 6 to 8 days after start of flight)
- if at peak flight (200 to 225 °D or 10 to 12 days after start of flight) the average number of moths/trap/day for any 3 to 4 day period is
  - — 5 or more, treat within 3 days of peak
  - — less than 5, no treatment necessary
- continue trap counts every 3 to 4 days until 2½ weeks after start of flight
- check traps weekly through August to see if a third flight appears

Moths of Second Generation (Third Flight)

No guidelines are available for the third flight. However, if you observe larval entry into nuts (frass at stem end), a treatment may be warranted.

Some insecticides are more destructive to the walnut aphid parasite and to predaceous mites than other materials. If you treat the first brood, choose a material that allows the aphid parasite to reestablish itself quickly. If you treat either brood, watch out for mite buildups from July to August. For recommendations see the latest University of California control guide on walnuts.

DATE	DAILY LOW	DAILY HIGH	DEGREE-DAYS FOR DATE	TOTAL DEGREE-DAYS FOR SEASON TO DATE
4/1 *	41 °F	61 °F	4	4
4/2	37	64	5	9
4/3	37	63	4	13
4/4	42	61	4	17
4/5	53	61	7	24
4/6	46	61	4	28
4/7	39	71	8	36
4/8	40	78	11	47

FIGURE 31. Record the daily high and low temperatures and the degree-days for the date. The total degree-days for the season to date will tell you when to expect each codling moth flight.

A "snout" formed by a pair of palps helps distinguish this moth from the codling moth.

The larvae consume nut meat and produce frass and substantial webbing; they change into pupae inside the nut.

Navel Orangeworm
Amyelois transitella

Although not as severe a pest in walnuts as in almonds, the navel orangeworm can cause substantial economic loss to walnuts when populations are allowed to build up in nuts damaged by codling moth or blight or in mummy and trash nuts in the orchards. It infests sound nutmeats of the new crop after husk split and is most often a pest of those cultivars that are also susceptible to codling moth and blight.

Description and Seasonal Development

The navel orangeworm overwinters in various stages of nondiapausing larvae inside mummy nuts on the tree and in trash nuts left on the ground and around hullers. Pupation begins in March and may continue through early May. The moths of the overwintered brood start emerging in April, and peak emergence usually occurs from late April to mid-May, depending on season and locality.

The adult moth is silver gray with irregular black patches on the forewings. The snoutlike palps in front of the head help distinguish this moth from the codling moth. Females of the overwintered generation lay their eggs singly on damaged nuts. They prefer nuts remaining on the trees, but if none are available, they will lay their eggs on nuts on the ground. The first generation, and most of the second, is completed in these nuts. In late summer, some of the second generation larvae infest the new crop as the husks begin to split. Females emerging at this time prefer to lay eggs on the opened husk or on the exposed nutshell. Larvae of the third generation begin entering the new crop about 30 days after start of husk split. Crops harvested before this time usually escape severe infestation and damage. Figure 32 shows a typical seasonal development of the navel orangeworm.

The eggs of the navel orangeworm are opaque white when first laid. After about a day, they turn pink, then reddish orange. The navel orangeworm larva has a pair of brown, crescent-shaped marks on the second segment behind the head. These marks are absent on the codling moth larvae. After hatching, the tiny caterpillars enter nuts through the soft tissue at the stem end and do not emerge until they are adults. Several larvae may infest one nut and produce substantial webbing. In contrast, only a single codling moth larva is found in each nut, and the codling moth produces little webbing.

Damage

Nuts infested with the navel orangeworm are unmarketable because the larvae feed on the nutmeats and produce webbing and frass. Navel orangeworms do not damage sound walnuts until the husks begin to split. Nuts infested only by the navel orangeworm show no external symptoms, so that nuts must be opened to detect an infestation.

Management Guidelines

Prevention. The navel orangeworm population in an orchard can be suppressed by good sanitation, by controlling codling moth and blight, by preventing sunburn and by harvesting as early as possible. Where possible, rid the trees of mummy nuts after harvest. Clean berms and disc, shred or burn trash nuts and huller waste materials before mid-March to destroy developing larvae and eliminate these food sources for the first generation larvae. Encourage neighbors that grow walnuts or almonds to practice good orchard sanitation; navel orangeworm moths fly in from adjacent infested orchards.

Harvest as soon as possible after husk split to shorten the period during which the nuts are exposed to attack. Using a plant growth regulator to advance husk split and harvest may be feasible where the navel orangeworm is a major problem. This practice can be helpful in preventing damage even when other preventive measures are neglected. Consult your farm advisor for the

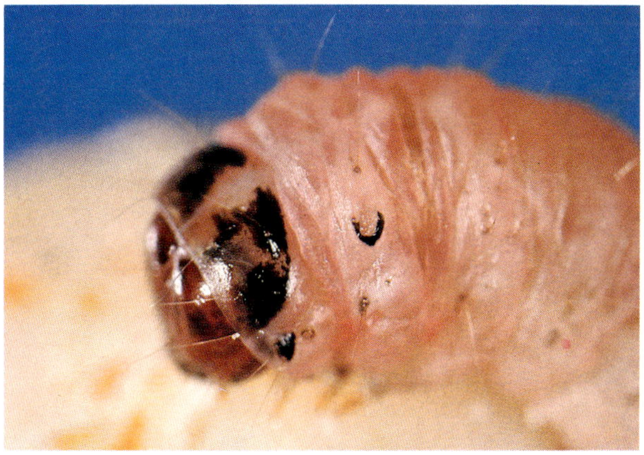

The larva has a pair of crescent-shaped marks on the second segment behind the head; the presence of these marks distinguishes it from the larva of the codling moth.

NAVEL ORANGEWORM

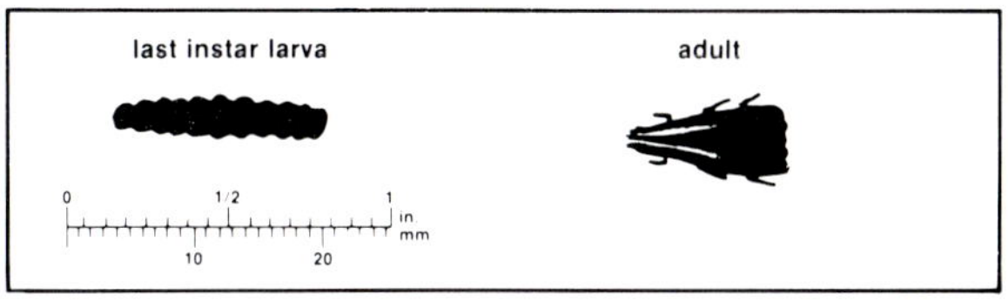

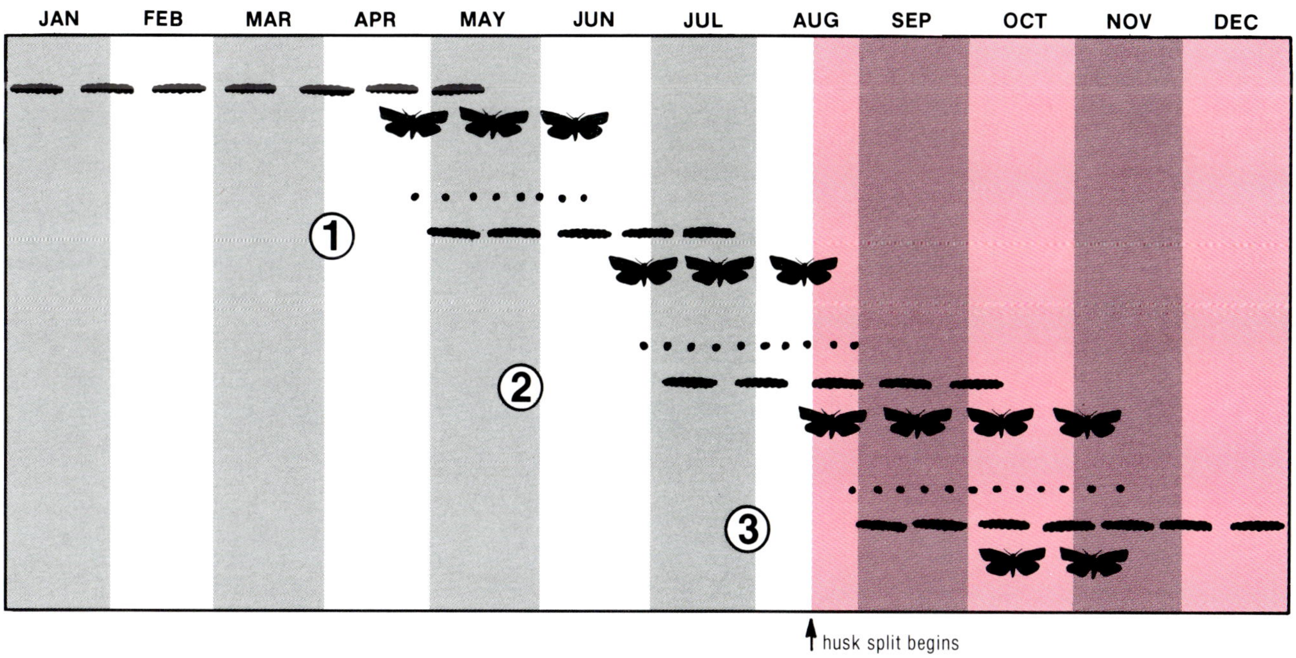

FIGURE 32. Seasonal development of the navel orangeworm. Appearance of broods and developmental times vary with temperature. Two full generations (1, 2) and a partial third generation (3) are produced during the season. A portion of the second generation and the third (overwintering) generation larvae enter sound nuts after husk split (red shaded area). Time of husk split varies with cultivar and temperature.

latest product and its application. After shaking, remove the crop from the orchard and hull and fumigate as soon as possible to prevent infestation. Early and rapid harvest also helps preserve nut quality.

A wasp that parasitizes the navel orangeworm, *Pentalitomastix plethoricus,* is established in many walnut and almond growing areas and accounts for some larval mortality. A ground cover maintained during the winter aids in the decomposing of trash nuts by molds and other microorganisms.

Monitoring and Control. There is currently no demonstrated method to monitor the navel orangeworm in walnut orchards. However, check your mummy nuts sometime in February for the navel orangeworm. Crack open the nuts and look for the larvae. If significant numbers of mummies are present, especially if a large proportion is infested, destroy the mummy and trash nuts. Insecticide applications against the navel orangeworm in walnuts are currently not recommended. Use the preventive tactics discussed above to contain your orchard population. Rapid harvest, hulling, drying and fumigation will suppress the development of the navel orangeworm in the harvested crop.

Walnut Husk Fly
Rhagoletis completa

The walnut husk fly is a mid- to late season pest. It occurs in all walnut-growing areas in California except in certain areas of the central and southern San Joaquin Valley. Black walnut and all cultivars of English walnut are suitable hosts for the husk fly. Some early maturing cultivars, such as Ashley, escape serious damage in most years. Mid-season and late-maturing cultivars, such as Eureka, Hartley and Franquette, are most susceptible to husk fly damage.

Description and Seasonal Development

The adult of the walnut husk fly is about the size of a house fly and very colorful. A yellow spot just below

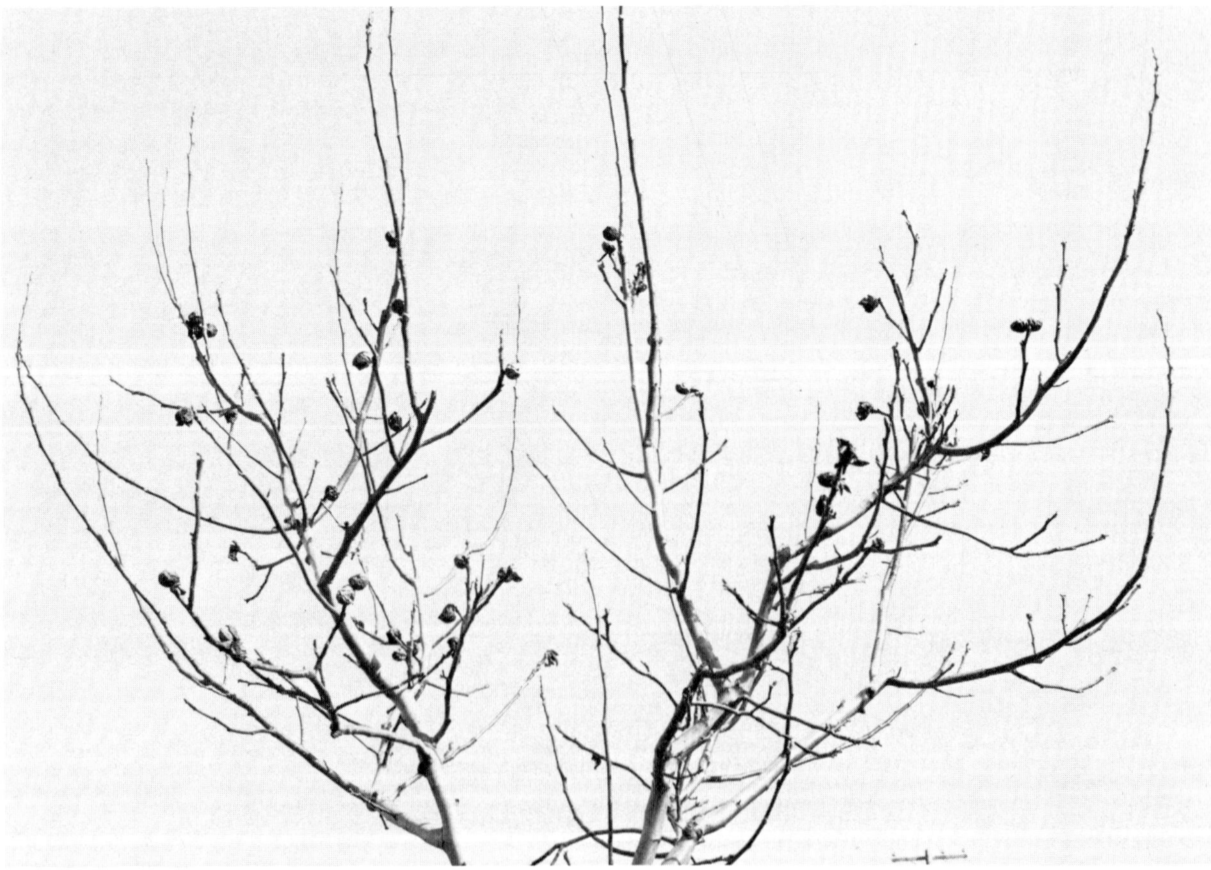

Mummy nuts provide overwintering sites for the navel orange worm.

the area where the wings are attached and a dark triangular band at the tip of the wings distinguishes the husk fly from other flies likely to be found in orchards.

Husk flies produce one generation per year. They overwinter as pupae in the soil and emerge as adults from early July until early September. Peak emergence is usually in mid-August. They mate about eight days after their emergence. Several days after mating, the female deposits eggs in groups of about 15 below the surface of the husk. The eggs hatch into white maggots within five days. Older maggots are yellow with black mouthparts. After feeding on the husk for three to five weeks, mature maggots drop to the ground and burrow several inches into the soil to pupate.. Most emerge as adults the following summer but some remain in the soil for two years or longer.

Damage

The first signs of an infestation are the small "stings" caused by females depositing eggs in the husk. Afte hatching, the maggots feed inside the husk, turning very soft and black. The outer skin of the husk usu remains intact, but its fleshy parts decay and stain nutshell. These stains cannot be removed by n bleaching procedures, and the nut is therefore u factory for in-shell sale. A husk fly infestation the season (late July to mid-August) leads to and darkened kernels or may induce mold gro damage may also be caused by other pests mental stresses. Late infestations do little da kernels but may stain the shells.

Management Guidelines

Prevention. Remove or treat aban walnut trees, especially black walnut favorite host. Native parasites and are not effective in controlling the

Monitoring and Control. M determine if flies or young maggots when to treat. Control action thresholds available.

Walnut husk fly adults can be monitored with various types of traps. In the past, the Frick trap and a trap baited with glycine and lye have been used. The Frick trap is no longer available, and the glycine/lye trap is troublesome to service.

At present, the recommended trap is the Pherocon® AM, which was developed for monitoring the apple maggot, a species closely related to the husk fly. This trap is a yellow, rectangular cardboard card coated with a mixture of protein hydrolysate and ammonium acetate to attract the flies and a sticky substance to trap them (Figure 25). As a bait trap, it catches both males and females of the husk fly and often other fly species too. Clean traps once or twice a week. The traps are effective throughout the season unless trap catches are very high or dust accumulates on the sticky surface.

To trap husk flies, use a trapping density similar to that recommended for the codling moth. At the beginning of July, hang traps at least 6 feet above the ground in a shady part of the tree. If husk flies were present the previous year, hang the traps in trees that were infested. Use the same trap trees year after year. It is a good practice to monitor routinely for husk fly even if it has not previously been a problem.

To determine whether an orchard needs to be treated for husk fly, review the history of husk fly damage in that orchard. If there was no significant damage due to the husk fly in the past 2 to 3 years, treatment may not be needed. If some damage occurred and trap counts indicate a gradual buildup, a treatment is probably warranted.

There are two options for treating the husk fly: con- ol of the adults or control of the hatching maggots. reating adults, spray 10 days after the number of as increased significantly. You can use an d with Staley's Protein Bait #7 to at- e insecticide. With this combination, er tree row is sufficient for immediate festation occurred early, you may have d time after 25 days if trap catches have n. However, treatment is not necessary ks have begun to split.

hoose to treat hatching larvae, monitor nuts ek for egg punctures or stings. Start monitor- two to three weeks after adults began to appear aps. Inspect a minimum of 10 nuts from shady f 20 vigorous trees throughout the orchard. Check uts closely for stings, especially at the stem end. stings will be visible just after egg laying. A dark e may flow from the egg puncture, leaving a charac- ristic tear stain. To confirm your diagnosis, check for husk fly eggs or maggots within the puncture.

Treat when the infestation is approaching 5% of the sampled nuts. To treat nuts with stings successfully, you will have to use a systemic insecticide that is more disruptive to predators and parasites of scales and mites than bait control for adults. This material is only effective against eggs and newly hatched maggots; it is not effective against maggots that have developed far enough to cause substantial darkening of the husk tissue. Waiting until egg punctures appear has the advantage that usually only one insecticide application is needed to control a population or that no application is needed if the stings appear after the husks have begun to split. For recommendations see the latest University of California control guide on walnuts.

A yellow spot on the midsection of the body and a triangular band at the tip of the wings distinguish the husk fly from other flies in walnut orchards.

If you remove the husk surface from a stung nut, you can see the eggs and sometimes newly hatched maggots in a small cavity.

HUSK FLY

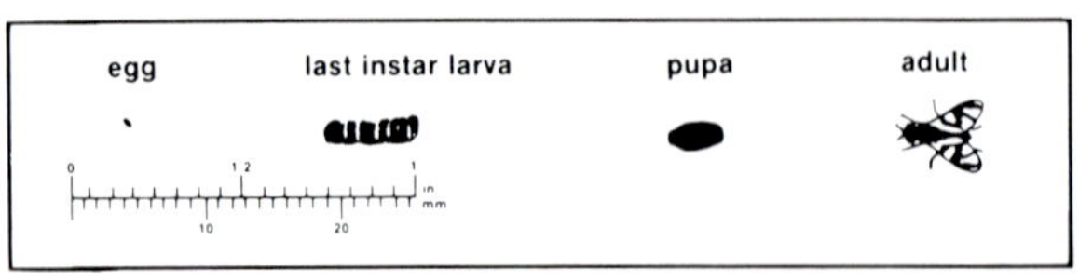

The egg puncture or "sting" is visible after the female has deposited her eggs just beneath the surface of the husk.

Maggots are whitish when young and become yellow as they mature.

Husks infected with husk fly have black, soft areas; eventually the entire husk may be affected.

Walnut Aphid
Chromaphis juglandicola

Dusky-Veined Aphid
Callaphis juglandis

The two aphid species that may damage walnut trees are the walnut aphid and the dusky-veined aphid. Their seasonal development is very similar, but their appearance and behavior are quite different. The walnut aphid used to be a major pest in walnuts, but it is now mostly controlled by an introduced parasitic wasp. Since the biological control of the walnut aphid, the dusky-veined aphid has become a pest in some orchards.

Adult female aphids are commonly winged during the spring and summer. The dusky-veined aphid (left) has dusky markings along the veins of the wings and is much bigger than the walnut aphid (right).

Description and Seasonal Development

You can easily distinguish walnut aphids from dusky-veined aphids. Walnut aphids are much smaller than dusky-veined aphids and are typically found scattered on the lower side of leaves; dusky-veined aphids feed in rows along the midvein on the upper surface. During the spring and summer, adult females are commonly winged; dusky-veined aphids have wings with distinctive dusky markings along the veins. The nymphs of the dusky-veined aphid have dark, banded spots on the back; these spots are much less pronounced or absent on the nymphs of the walnut aphid.

Walnut aphids feed on the lower side of leaflets (right); dusky-veined aphids feed in rows along the midrib on the upper side (left).

The life cycle of the two species is basically the same. Both aphids overwinter in the egg stage on the twigs. The eggs hatch into stem mothers as soon as the leaf buds of early cultivars begin to open. The stem mothers settle on the leaflets and reproduce without mating, giving birth to live nymphs. The aphids pass through many generations a year, depending upon temperature. Aphids are most damaging in spring and early summer when rapidly growing nuts are particularly susceptible to stress. Temperatures of 38 °C (100 °F) or more maintained over several days in the orchard suppress dusky-veined aphid populations, whereas the walnut aphid is less susceptible to heat. The walnut aphid parasite apparently is more sensitive to heat than the walnut aphid, and in some summers, damaging populations of walnut aphids may develop. In the fall, wingless females and smaller, winged males appear. After mating, the females lay eggs, which overwinter.

Walnut aphid nymphs do not have distinct markings on their backs (right). Here one nymph was parasitized by *Trioxys*; the wasp develops inside the aphid and leaves the mummy through an exit hole. Nymphs of the dusky-veined aphid have dark banded spots on their backs (left). Newly molted individuals are bright yellow (left).

DUSKY-VEINED APHID

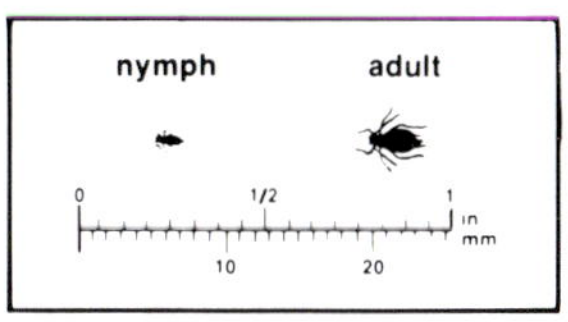

WALNUT APHID

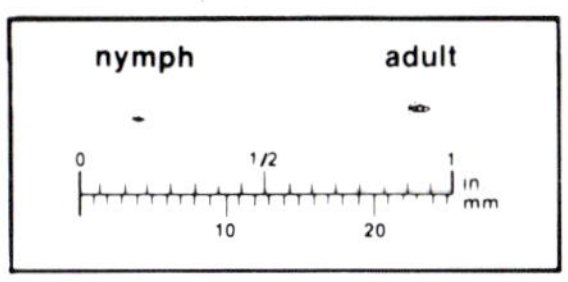

Damage

Aphid feeding can reduce tree vigor, nut size, yield and quality. Aphids excrete honeydew, which is toxic to the husk of some cultivars and turns the husk surface black. Sooty mold also develops on the honeydew on all cultivars, increasing the chance for sunburn on exposed nuts. High populations of aphids may lead to leaf drop, exposing more nuts to sunburn, which darkens or shrivels the kernels.

Walnut aphid populations of over 15 aphids per leaflet early in the season reduce nut yield and quality and cause an increase in nuts with perforated shells. An infestation in summer lowers the nut quality. Some late cultivars, such as Franquette, may tolerate heavy populations.

The feeding of dusky-veined aphids causes the midribs of leaves to turn black. A correlation has been established between infestation of dusky-veined aphids and nut quality. If 10 to 15% of the leaflets are colonized for 3 to 4 weeks before shell hardening, nut size is decreased. The same level of infestation during late summer will result in shriveled kernels at harvest time.

Management Guidelines

Prevention. The key to controlling the walnut aphid is to preserve the small parasitic wasp, *Trioxys pallidus*. This wasp, which was introduced from France and Iran in the 1960s, has virtually eliminated the walnut aphid as a pest in most orchards. The female lays an egg inside walnut aphid nymphs. The egg hatches into a larva, which consumes the inside of the aphid and pupates in the "mummified" aphid. Adult wasps emerge from the mummy through a small exit hole and parasitize more aphids.

Predatory insects also play a role in controlling both the walnut aphid and the dusky-veined aphid. Two lady beetles are important predators: the familiar convergent lady beetle, *Hippodamia convergens*, and the ashy gray lady beetle, *Olla abdominalis*, which is often numerous

ORCHARD LOCATION ______________ MONITORING PERIOD ________ to ________

BLOCK NUMBER ______________ INSECTICIDE TREATMENTS ______________

DATE	AVERAGE NUMBER OF APHIDS PER LEAFLET PER TREE																				TOTAL	AVERAGE	NOTES (predators, etc.)
	1	2	3	4	5	6	7	8	9	10	11	12	13	14	15	16	17	18	19	20			

FIGURE 33. Keep monitoring records for the walnut aphid in a form such as this one. To obtain the average number of aphids per leaflet per tree, count the total number of aphids on 5 leaflets and divide this total by 5. Repeat this method on 20 trees in each block of about 20 acres. Use the average number of aphids per leaflet throughout the block to make treatment decisions. Also note the presence of *Trioxys* and predators.

but not as noticeable because it spends all its time in the trees. Green lacewings, *Chrysopa carnea* and *C. nigricornus*, are common in walnut orchards; the larvae of both lacewings and the adult of *C. nigricornus* feed on aphids. Brown lacewings, *Hemerobius* spp., prey on aphids as well but usually are not as abundant in walnut orchards as green lacewings. You may spot a number of other aphid predators in the orchard, including syrphid flies and minute pirate bugs. Fungal diseases also infect and kill aphids, but their impact on populations is unknown. Aphids killed by fungus turn red and eventually shrivel.

Insecticide treatments can upset aphid control by killing natural enemies, especially the walnut aphid parasite. When this occurs, damaging populations of aphids develop rapidly. Watch aphid populations carefully after treating orchards or adjoining crops or fields, particularly in the spring when conditions are most favorable for aphid development. Delaying treatment for the codling moth until the second brood will help preserve natural enemies of aphids in the spring.

Monitoring and Control. Monitor walnut aphids and dusky-veined aphids by counting the number of aphids on leaflet samples. Begin sampling in May; sample every other week until populations start to build up, then sample weekly through August.

Collect your samples by picking mature leaflets; avoid leaflets from a new flush of growth because they may be heavily infested and not typical of the entire orchard. Check the upper surface of each leaflet for dusky-veined aphids and the lower surface for walnut aphids.

Take 5 leaflets from each of 20 trees per 20 acres. Choose trees that are in the interior parts of the orchard. Sample the edges of the orchard independently because drifting insecticides sometimes disrupt natural control in these trees, making border treatments necessary.

Count the walnut aphids on each leaflet and record the numbers in a format such as given in Figure 33. Also look for walnut aphid mummies; these indicate that the parasite is present in the orchard and may control the population. If you find an average of 10 nonparasitized walnut aphids per leaflet, sample again in about four days. If the average number of healthy walnut aphids rises to 15 per leaflet early in the season, the parasite is not effective and a treatment is warranted in early leafing cultivars. If you detect dusky-veined aphids in your orchard, watch closely for predators to see if they will control the aphids. If the aphid population increases to the point that 10% of the sampled leaflets have a colony of six or more healthy adults and nymphs, a treatment is necessary.

Choose insecticides that are least disruptive to the biological control of mites and scales and that allow the aphid parasite to recover quickly. For recommendations see the latest University of California control guide on walnuts.

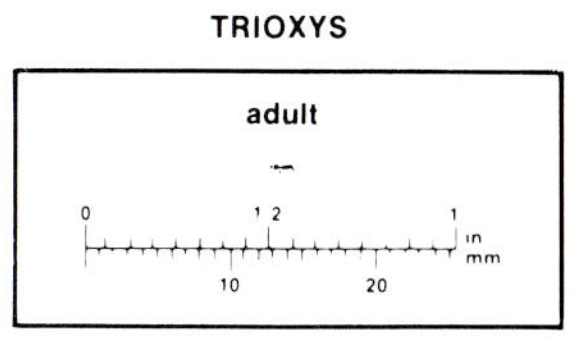

The walnut aphid parasite, *Trioxys pallidus*, inserts an egg into a small nymph.

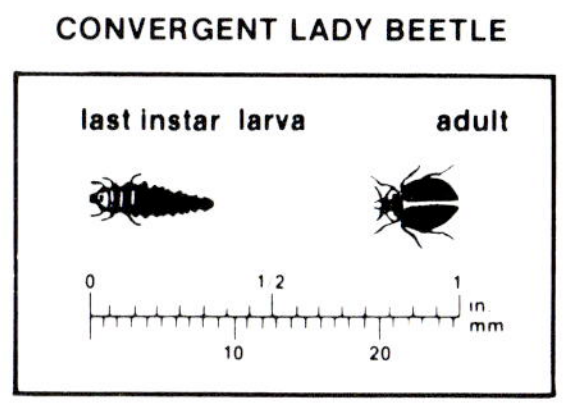

Lady beetles are important predators of aphids. Their larvae and eggs look very similar. Larva of the convergent lady beetle (above); ashy gray lady beetle (below, left); eggs and recently hatched larva of the ashy gray lady beetle (below, right).

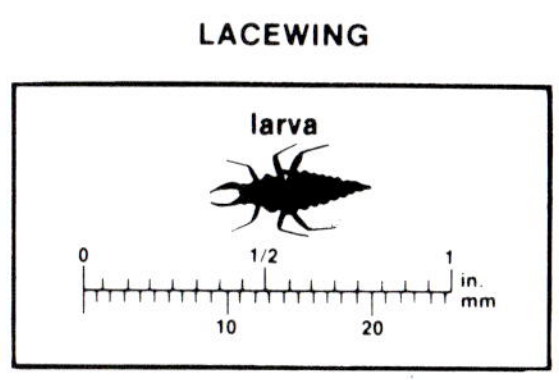

The larvae of lacewings feed on aphids, other small insects and insect eggs; larva (left) and egg (right) of a green lacewing.

The cover of the adult female of the walnut scale is circular and varies in color from gray to brown. The nipple is not as pronounced as in the San Jose scale.

The body of the female walnut scale has indented margins; here the covers have been removed to show this character. The upper right of the photo also shows three young scales in the white cap stage.

Walnut scales often form groups resembling a daisy; this formation results when male crawlers settle under the margin of the female cover and form their elongated cover.

WALNUT SCALE

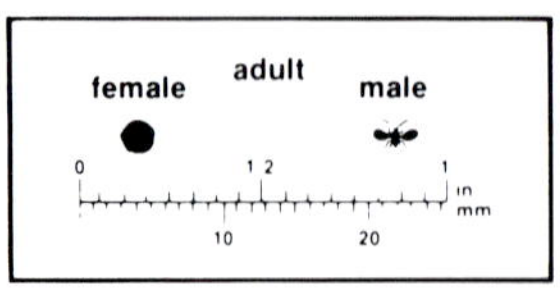

SCALE INSECTS

Both armored scales and soft scales occur in California walnut orchards, but the armored scales are the most damaging. Several characteristics distinguish the armored scales from the soft scales. The cover of the armored scale is separate from the body of the insect, whereas the cover of the soft scale is the actual body wall of the insect and cannot be removed. Unlike soft scales, armored scales do not secrete honeydew.

Among the armored scales, only the walnut scale and the San Jose scale often occur at damaging levels. The Italian pear and oystershell scale are occasional pests. Although once a significant pest problem, the Putnam scale (*Diaspidiotus ancylus*) today is rarely seen in walnut orchards. Several other species of armored scales have been reported on walnuts, including the fig scale (*Lepidosaphes conchiformis*), the olive scale (*Parlatoria oleae*) and the Buckeye scale (*Diaspidiotus aesculi*).

The most important soft scale on walnuts is the frosted scale; the European fruit lecanium and the calico scale are occasionally found but are seldom pests.

Walnut Scale
Quadraspidiotus juglansregiae

The walnut scale was considered a minor pest of walnuts until the 1960s when populations developed to significant levels in all major walnut growing areas of the state. The monitoring and management guidelines given below for this pest also apply to the other armored scales.

Description and Seasonal Development

The cover of the walnut scale varies from gray to brown and is often overlooked because it blends well with the color of the walnut bark. This scale insect is often found in daisy-shaped groups; this formation results when male crawlers settle under the margin of the female cover and begin forming their elongated covers. The cover of the adult female is circular. The body of the female is yellowish and has typical indented margins; these help distinguish the walnut scale from other armored scales on walnuts.

The walnut scale produces two generations a year in the Central Valley. The second generation overwinters as second instar females and males. In early spring, the females molt once more to become adults and then remain sessile, whereas the males molt three more times while developing eyes, antennae, legs and wings. Both sexes mature at the same time. After mating, the females lay eggs in mid-May. When first laid, the eggs sometimes stick together, forming a chain. They hatch in two to three days. The young female scales or crawlers move around the branches for a short time before they settle down, begin feeding and secreting the scale cover. Male crawlers usually move only to the margins of the female cover. Initially, the scale cover is white; this is called the white cap stage. The white caps change to gray or brown covers after about a week. This first generation completes its development by mid-July, and the females lay eggs in mid-August. The crawlers of the second generation molt once before they overwinter.

During the fourth instar or pupal stage of the walnut scale male (left), the antennae, eyes, wings and penal sheath are distinguishable. The adult male (right) has well-developed antennae, eyes, legs and one pair of wings.

Damage

Armored scales suck plant juices from the inner bark by inserting their mouth parts into twigs and branches. Infested trees look water-stressed, and inside fruiting wood on lateral bearing cultivars may die back when encrusted with scale insects. Extremely heavy populations can cause the bark to crack.

A heavy infestation of walnut scale may encrust small limbs and twigs, eventually killing them.

Management Guidelines

Prevention. Several natural enemies help control the armored scales. Two predaceous beetles, the twice-stabbed lady beetle, *Chilocorus orbus*, and another small beetle, *Cybocephalus californicus*, often feed on scale insects. A number of small wasps parasitize armored scales. The most important are *Aphytis* spp., during the fall and *Prospaltella* spp. during the spring and early summer. These predators and parasites can be very effective in controlling scales. However, insecticides used for key pests can disrupt this natural control.

TWICE-STABBED LADY BEETLE

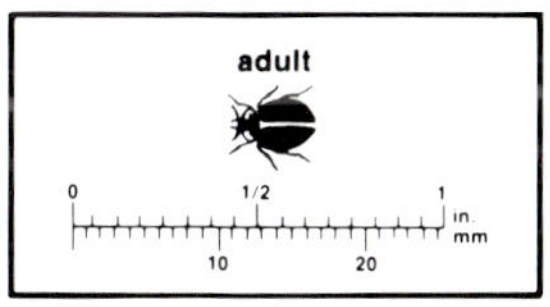

CYBOCEPHALUS

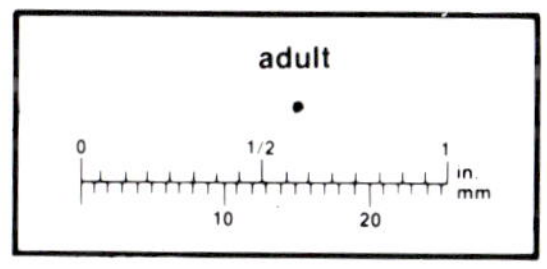

Monitoring and Control. At present, there are no numerical guidelines indicating when walnut scale populations require chemical control. Each orchard has to be judged individually, taking into account the current scale population and the history of the orchard.

Always monitor the orchard to determine if insects are present; dieback of fruiting spurs or cracked bark may indicate a scale infestation but may also reflect other stresses. If you find scale insects in moderate to heavy numbers, treat the orchard.

The twice-stabbed lady beetle feeds on armored scales (left). A smaller beetle, *Cybocephalus* sp., also attacks armored scales (right).

The best time for monitoring and control is the dormant or delayed dormant period when all parts of the tree, especially the upper limbs, can easily be checked by examining the prunings. Walnut scales clump in small areas but are difficult to see because the color of their covers blends well with the bark.

Also closely examine the trunk and scaffolds because walnut scales usually start in the lower limbs when they infest a tree. Scale populations tend to be spotty, so check the entire orchard at least once a year. Often only a spot treatment is needed for control.

Aphytis spp. are tiny yellow wasps that parasitize armored scales.

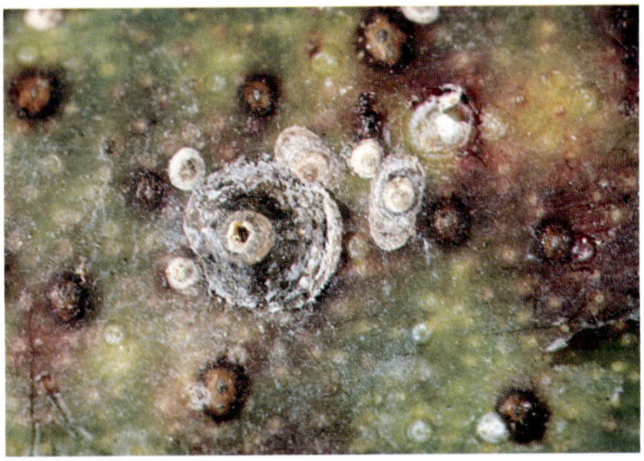

The cover of the female San Jose scale is circular and has a distinctive nipple, which is surrounded by a black or bluish band; the immature male cover is elongated.

If you remove the cover, you can see that the body of the female San Jose scale is not indented, as is the body of the walnut scale.

The San Jose scale overwinters as first instar nymphs in the "black cap" stage.

SAN JOSE SCALE

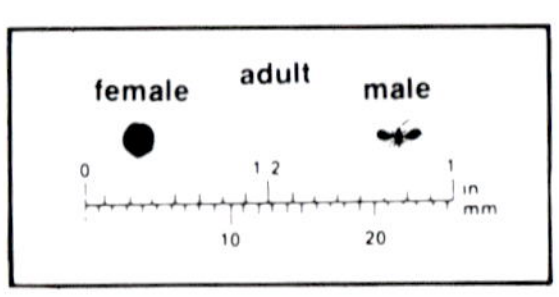

The San Jose male has a dark band across its back.

Treat the walnut scale during the delayed dormant period after male scales begin emerging. Check under the elongated male covers for pupal stages; the adults emerge about three days after pupation. Emergence usually occurs just before catkin elongation on early cultivars. At this time, you can achieve good coverage of limbs and scaffolds and avoid killing natural enemies. Often a combination of an insecticide and oil is used. However, do not use oil sprays in orchards that have been stressed by drought, heavy scale populations or other factors during the growing season or the dormant period.

You can also monitor walnut scales in late May or early June when they are in the "white cap" stage. Treatment at this time is not recommended because it kills the natural enemies of aphids and mites. For recommendations see the latest University of California control guide on walnuts.

San Jose Scale
Quadraspidiotus perniciosus

The San Jose scale attacks a wide range of woody plants and is a pest of deciduous trees and ornamentals in many parts of the world. Since the 1940s, this scale has become numerous in some California walnut orchards.

Description and Seasonal Development

The covers and body of the San Jose scale are similar to those of the walnut scale, except the margin of the female body of the San Jose scale is not indented. The male of the San Jose scale looks similar to the male of the walnut scale but has a black band across its back.

The San Jose scale produces three generations a year or more if warm weather extends well into the fall. It overwinters mainly as first instar nymphs in the "black cap" stage. The female nymphs molt twice before reaching maturity in March. The male nymphs molt four times and mature at the same time as the females. The developmental stages are similar to those of the walnut scale but the San Jose scale bears live young. The tiny crawlers begin emerging in May, and they soon settle down and secrete white waxy covers, the "white caps." After about a week, the scales begin secreting a dark wax, forming the black caps. After two to three weeks, the nymphs molt. After this molt, males and females become distinguishable and complete their development in three to four weeks or after 400 to 500 degree-days.

The generations produced during the summer overlap and crawlers are present throughout the summer and fall. The mixed brood that builds up in late fall overwinters as first instar black caps, although some mature, mated females may be present.

Damage

The damage caused by San Jose scale is similar to that caused by the walnut scale. The infested bark often cracks and dies, and heavily infested scaffold limbs and branches die within one to two years.

Management Guidelines

Prevention. The predators and parasites listed for the walnut scale also attack the San Jose scale and play a significant role in keeping scale populations below economically damaging levels. Where heavy populations exist, however, these beneficial species may not be able to control the scale insects before significant damage has occurred. When insecticides applied against other pests have disrupted the natural control, populations often increase dramatically. Low mortality due to mild winter temperatures will also permit a buildup of scale populations.

Monitoring and Control. As with the walnut scale, the preferred time for monitoring and treating San Jose scale is the dormant or delayed dormant period. Examine scaffold limbs, branches and prunings for the characteristic black caps and old scale bodies. If a population needs treatment, an application at this time can achieve better coverage and control and avoids killing natural enemies.

You can also monitor the San Jose scale in the spring by using pheromone traps. Place the traps in the orchard in early March to detect the first emergence of males. Methods are being developed using the male flight to forecast the emergence of crawlers in May. In mid- to late May, you can also look for the white or black cap stages.

If a heavy population requires an additional treatment during the season, it should be applied against crawlers or white and black caps; later stages are extremely difficult to control. Treatment at this time, however, is very disruptive to the natural enemies of aphids and mites.

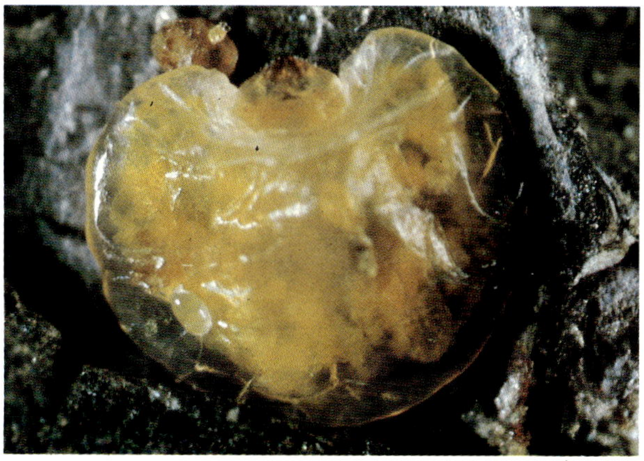

A parasitic wasp (*Aphytis*) has laid its egg on the lower left side of a female scale.

An *Aphytis* pupa with the remains of the scale body on the left. The black pellets surrounding the pupa are the meconia excreted before pupation.

Italian Pear Scale

Epidiaspis leperii

The Italian pear scale is widely distributed in walnut orchards in California but builds up to heavy populations only when moss and lichen are allowed to cover trunks and larger limbs. The female body is pink or reddish, a trait that helps distinguish the Italian pear scale from other armored scales found in walnut orchards.

Light to moderate infestations do not seem to harm trees. Heavy scale aggregations may cause the bark to crack and can reduce tree vigor. Such large populations,

Italian pear scales are often found on trunks and limbs that are covered with lichens. Removing the scale covers reveals the pink or reddish body of the female scale. The color helps distinguish this scale from others found on walnuts.

ITALIAN PEAR SCALE

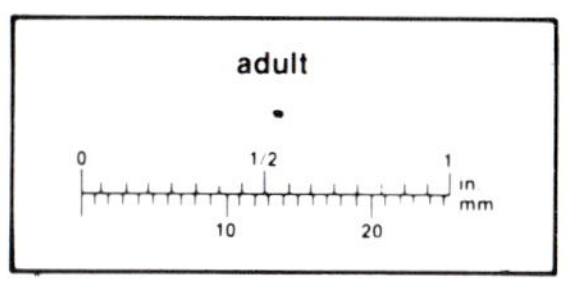

however, are seldom encountered in orchards that are regularly treated for blight.

Look for the Italian pear scale during the dormant period when you monitor other scale insects. Check especially scaffold limbs and branches that are covered with lichens. Scrape the lichens to look for the gray scale covers and the pink scale bodies underneath. The key to managing Italian pear scale is the control of lichens. If you find scale insects, remove the moss and lichens. Blight treatment in the spring will control the growth of moss and lichens. For recommendations, see the latest University of California control guide on walnuts.

OYSTERSHELL SCALE

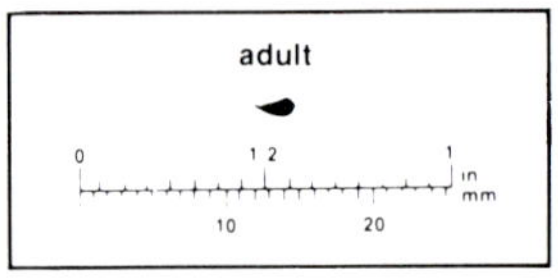

The cover of the oystershell scale resembles a mussel or oyster.

Mature frosted scales have convex, shiny brown covers. Parasitized adults are easily recognized by the exit holes, from which mature parasites have emerged.

W. GELLING

Frosted scales overwinter as small, flat, brownish nymphs.

FROSTED SCALE

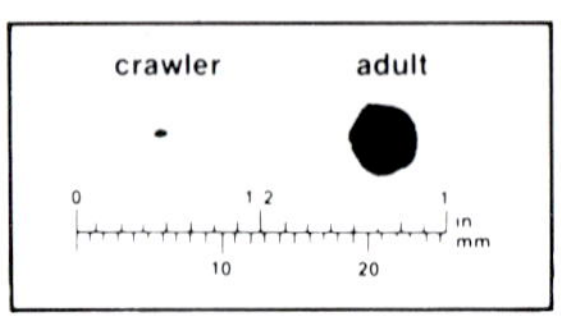

Oystershell Scale
Lepidosaphes ulmi

The oystershell scale is seldom a pest. Outbreaks usually occur near Lombardy poplar trees, which are its favorite host. This scale insect produces only one generation a year in California. It feeds on bark, causing leaves to turn yellow and dry up. Heavy populations will kill small limbs and occasionally large limbs and scaffolds. Little is known about natural enemies of this scale.

If walnuts near Lombardy poplar trees show clusters of wilting or dessicated leaves, inspect limbs and scaffolds for the characteristic mussel-shaped scale covers. Oystershell scales start to build up on a single limb, so you can usually remove the problem by pruning out the infested limb. If you encounter a heavy population on more than one limb, a spot treatment with an insecticide may be necessary. Treatment is effective only in late May or early June when crawlers are present; dormant sprays are not effective against this scale because it overwinters in the egg stage under the old female covers.

Frosted Scale
Lecanium pruinosum

The frosted scale is the most important soft scale pest of walnuts. Its frostlike cover and copius honeydew secretion make it easy to recognize in the spring. Management guidelines for the frosted scale also apply to other soft scales attacking walnuts.

Description and Seasonal Development

The frosted scale produces one generation a year. It overwinters as nymphs on twigs and small branches. In the spring, it grows rapidly, becomes convex and forms a frostlike waxy cover. The scales secrete large amounts of honeydew at this time. After mating in late spring, the females lay a large number of eggs, which fill the entire space beneath the covers. After egg production, the females die and the white waxy substance weathers away, leaving an oval, dark brown cover that may be present for a year or more.

The newly hatched nymphs or crawlers emerge from beneath the scale from late May through June and settle mostly on the underside of leaves. Here they feed for the rest of the summer. In the fall, the nymphs molt and move back to twigs.

Damage

The nymphs of the soft scales suck plant juices from the leaves and twigs. Low to moderate populations apparently are not damaging, but heavy populations reduce terminal growth and vigor, resulting in smaller nuts and poor kernel quality. The secreted honeydew may cover nuts and favor the growth of sooty mold, increasing the chances for sunburn damage.

Management Guidelines

Prevention. Parasites play an important role in controlling soft scales. The natural control, however, can be disrupted by adverse weather conditions or by insecticides. Some materials may also provide adequate control of the scale insects.

The most effective parasite of the frosted scale is a wasp, *Metaphycus californicus*. It produces several generations a year, compared with one generation of the scale, and parasitizes all stages of the frosted scale except the eggs. Parasitized nymphs are almost black and have convex covers; unparasitized nymphs are flat. Several parasites emerge from a single parasitized adult scale, leaving a "perforated" cover. Look for these signs of parasitism because heavily parasitized populations usually need no further control.

Monitoring and Control. Monitor frosted scales during the dormant period. Examine the previous season's growth on randomly selected trees throughout the orchard. If you find more than five nymphs per foot of last year's wood throughout the orchard and less than 90% are parasitized, a treatment is warranted. A dormant spray can achieve the best coverage and control with the least destruction of beneficials. Treat no later than mid-January because adult scale insects, which start appearing at the beginning of February, are extremely difficult to control.

You can also monitor the frosted scale in spring and summer. From early March to mid-April, look for the powdery white wax that covers the adult scale insects. After this time, the wax erodes and the scale insects appear dark brown. At the beginning of July you can detect the crawlers on the underside of leaflets with a hand lens. Avoid a summer treatment directed against the frosted scale, as it may disrupt the natural control of aphids and mites. For recommendations, see the latest University of California control guide on walnuts.

Nymphs parasitized by *Metaphycus* are convex and almost black.

In the spring, the overwintered frosted scale nymphs start forming white, waxy covers, the "white caps."

This frosted scale cover has been removed to show the newly hatched nymphs or crawlers in early summer.

European Fruit Lecanium
Lecanium corni

The European fruit lecanium has essentially the same life cycle as the frosted scale. The immature stages closely resemble those of the frosted scale, but the adults do not form the thick, frostlike cover in the spring.

Use the same monitoring and control measures as described for the frosted scale.

TWO SPOTTED/PACIFIC MITE

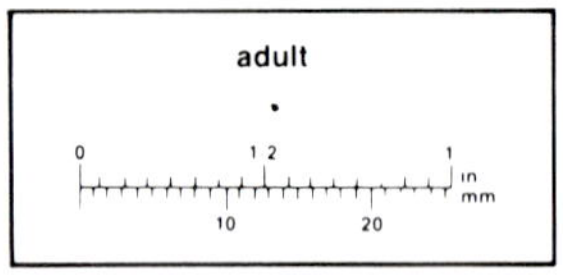

Pacific and two-spotted spider mites vary in appearance; the dark spots may barely be visible or may coalesce to large dark areas, making it difficult to distinguish the two species.

The spider mites overwinter as reddish orange, mature females. In the spring, they lay eggs, which are almost translucent but become opaque shortly before hatching.

Feeding by two-spotted or Pacific mites causes stippling and browning of leaves.

Calico Scale
Lecanium cerasorum

Although its life cycle is similar, the calico scale is different in appearance from both the frosted scale and the European fruit lecanium. Its cover has a typical white and brown "calico" pattern. The nymphs are larger and more elongated than those of the frosted scale and are covered with a thick, patterned coat in the summer and fall.

Use the same monitoring and control measures as described for the frosted scale.

Two-Spotted Mite
Tetranychus urticae

Pacific Mite
Tetranychus pacificus

Plant-feeding spider mites are a frequent problem because of the destruction of natural enemies and dusty conditions in the orchard. The two species most often found are the two-spotted mite and the Pacific mite.

Description and Seasonal Development

Pacific and two-spotted mites overwinter as reddish orange, mature females in protected places on the tree and in trash on the ground. During warm weather in spring, the overwintered females begin feeding on walnut leaves and ground cover in the orchard.

Colonies develop on the underside of leaves and also on the upper sides when heavy populations build up. The eggs are spherical and translucent when first laid, becoming opaque soon before hatching. Immature mites molt three times before becoming adults. The first stage mites have six legs; later stages and adults have eight legs. These mites reproduce rapidly in hot weather and may become numerous in June or July. They produce many generations a year. If temperature and food supply are favorable, a generation can be completed in seven days.

During periods of active feeding two-spotted mites have a dark spot on each side of the body; Pacific mites have a second pair of dark spots near the posterior end. Often, however, the spots are barely visible or may coalesce to large dark areas, making it difficult to distinguish the two species.

Damage

Mite feeding causes stippling and browning of leaves. Clusters of brown leaves are often the first sign of an increasing mite population. Heavy populations produce copious webbing, and their feeding causes leaves to desiccate and drop. Defoliation early in the season will greatly reduce nut yield and quality.

Management Guidelines

Prevention. Orchard management practices can reduce mite problems. Hot, dusty conditions often lead to mite outbreaks. You can minimize dust by oiling orchard roads and maintaining a ground cover. A ground cover also provides a cooler orchard. Well-irrigated, vigorous trees are less troubled by mite infestations.

Predators can keep spider mites below economically damaging levels. Avoiding dust and choosing selective materials when controlling other pests will help preserve predator populations. The most dependable mite predator is the western predatory mite (*Typhlodromus occidentalis*). Under optimal conditions, this predatory mite can produce a generation in seven days, allowing it to build rapidly and in many cases control the plant-feeding mites. Because walnut orchards early in the season often lack an alternate food source, such as European red mites, the western predatory mite may be late in building to numbers sufficient to control the Pacific and two-spotted mites, which become abundant later in the season.

Six-spotted thrips can be very effective in reducing high populations of the Pacific and two-spotted mites. Thrips, however, usually do not move into orchards until mite populations are high. Both the adults and the small, yellowish larvae prey on mites.

The spider mite destroyer (*Stethorus picipes*) is a small lady beetle that feeds on mites. The larvae are small, dull black and have a velvety appearance. The spider mite destroyer, like the six-spotted thrips, generally does not become numerous until spider mite populations are very heavy. But it is an active feeder and can reduce mite populations quickly when it is abundant. Minute pirate bugs also feed on spider mites.

Monitoring and Control. Begin looking for Pacific and two-spotted mites in walnut orchards in June or early July and continue checking once a week through August. If field bindweed is present, check it for mites. Mites will often colonize bindweed 10 to 14 days before moving into trees. Examine the leaves in "hot spots" and at the edges of the orchard where mites may be more numerous because of dust. Also look for brown clusters of mite-damaged leaves throughout the orchard. Colonies usually begin on the lower branches in the inner canopy. Pick infested leaflets at random and check for the presence of predators. If brown clusters of leaves are present on 10% of the trees and no predators are present, treat the orchard for mites. If predaceous mites, six-spotted thrips or spider mite destroyers are present

The most dependable mite predator is the western predatory mite. This photo shows the normal translucent appearance of the adult.

THRIPS

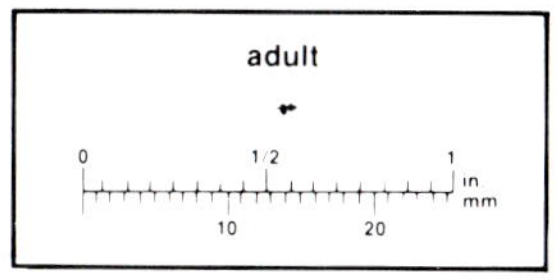

Six-spotted thrips are characterized by three dark spots on each forewing. They are effective mite predators but move into orchards only when mite populations are high.

The spider mite destroyer also preys on mites but becomes numerous only when mite populations are heavy; adult (left), larva (right).

SPIDER MITE DESTROYER

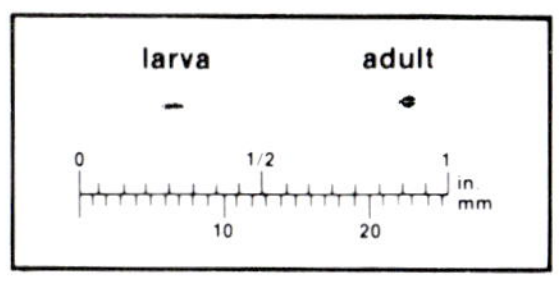

MINUTE PIRATE BUG

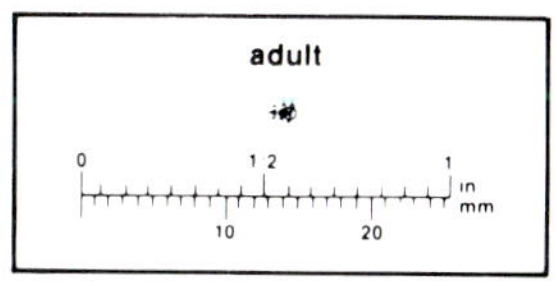

The minute pirate bug feeds on spider mites, immature scales, other small insects, insect eggs and newly hatched larvae.

Adult females of the European red mite have a brick red, globular body with four rows of long hairs; males are brownish and smaller.

The eggs of the European red mite are red and have a characteristic spike.

European red mites serve as an alternate food source for the western predatory mite. The predator turns reddish when feeding on this mite.

on half of the leaflets, natural enemies will probably control the population. Check again in 3 to 4 days to determine whether the plant-feeding mite population is building or declining. In very warm weather, the mites may build up faster than the predator populations.

If the population is increasing in June or July and it appears that leaf drop will be severe enough to expose the nuts to sunburn or to interfere with the sweeping operation at harvest, treat for mites. If mites do not build up in walnuts until the end of August, a treatment is seldom warranted. For recommendations see the latest University of California control guide on walnuts.

European Red Mite

Panonychus ulmi

The European red mite thrives under high humidity, moderate temperatures and shady conditions. It is primarily a pest in the coastal areas and only occasionally builds to damaging populations in the Central Valley.

Description and Seasonal Development

The European red mite overwinters in the egg stage on twigs and branches. The red orange eggs have a long spike that is visible with a hand lens. The eggs hatch in early spring when the walnuts leaf out, and the mites produce many generations during the summer. Immature mites are bright red; adult females have a brick red, globular body with four rows of long, curved hairs arising from white dorsal spots. The males are brownish and smaller than the females. Newly molted mites may appear greenish. Usually the red mite population builds slowly during the spring and does not become damaging until summer. European red mites produce little or no webbing.

Damage

European red mites feed on the cell contents of leaf tissue. Initially, the feeding causes light stippling of the leaves. Prolonged feeding by a heavy population will gradually give the leaves a bronzed appearance. Feeding by the European red mite does not result in leaf drop, as does feeding by the two-spotted and Pacific mite, but severe, repeated bronzing of leaves can reduce nut yields significantly.

Management Guidelines

Prevention. In low numbers, the European red mite can be beneficial by providing a food source for the western predatory mite (*Typhlodromus occidentalis*) in the spring. This predator can build its population to numbers that may be sufficient to control two-spotted and Pacific mites, which appear later in the season. The other predators described in the section on two-spotted and Pacific mites also attack the European red mite.

Monitoring and Control. Monitor European red mites in the winter and summer. During the dormant period, examine leaf and growth scars on twigs for clusters of the red eggs. During the summer, look for stippling or bronzing on leaves, especially in the shady, central parts of the trees. You can check for European red mites when sampling for two-spotted and Pacific mites. No damage threshold levels are available to determine when to treat. Avoid treating low to moderate levels of European red mites because they can be important in maintaining predators of other mites.

Redhumped Caterpillar
Schizura concinna

The redhumped caterpillar occasionally becomes a pest of walnuts in the warmer walnut-producing districts. This pest feeds on a wide range of deciduous trees, especially plums, prunes, apples, apricots, and on liquid amber and other ornamentals.

The redhumped caterpillar produces three generations per year. The brown moths of the first generation emerge in early May. After mating, the females lay pearly white, spherical eggs in masses of 25 to 100 on the underside of leaves. The young larvae feed in large groups, skeletonizing leaves. Heavy infestations can defoliate trees, especially in young orchards. After pupating on the ground, second generation moths emerge in July. Larvae of this generation develop into adults that produce a third generation in September.

A number of natural enemies attack redhumped caterpillars, frequently preventing them from becoming destructive pests. Most common are two parasitic wasps, *Hyposoter fugitivus* and a species of *Apanteles*. The larvae of both parasites develop inside the caterpillar and pupate on the leaf surface. Several general predators, including spiders, lacewings, bigeyed bugs, and damsel bugs, occasionally feed on caterpillar eggs and small larvae.

Monitor throughout the orchard in May for the first brood of caterpillars. Check for colonies of young larvae on the underside of leaves, where their feeding will give the leaves a translucent or net-like appearance. Later on, the larvae will consume entire leaves except for the midrib and major veins. If you detect an infestation, do not treat for the first brood. Wait for the second brood to emerge in July before deciding to treat.

If single, young trees are infested, spot treat with a hand spray gun. For large trees or areas, either ground or

The eggs of the redhumped caterpillar can be found in clusters of 25 to 100 on the underside of leaflets; the hatched larvae feed in groups.

Mature larvae are very colorful and have a characteristic red hump (left). The larvae feed on the walnut leaf tissue, leaving only the midrib and larger veins (right).

Several parasites attack the redhumped caterpillar. *Apanteles* develops inside the caterpillar and pupates in groups of silken cocoons.

RED-HUMPED CATERPILLAR

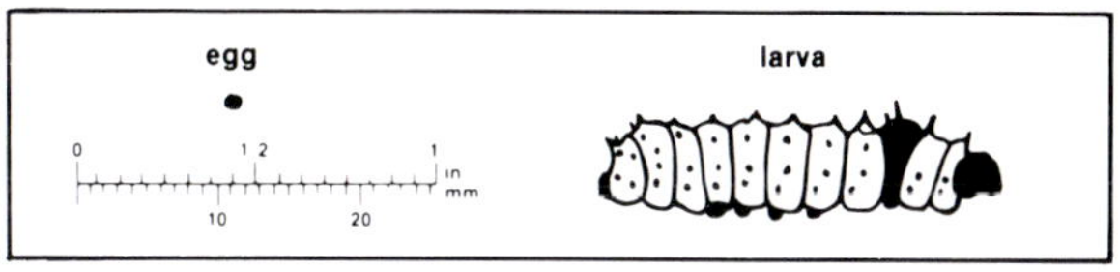

FALL WEBWORM

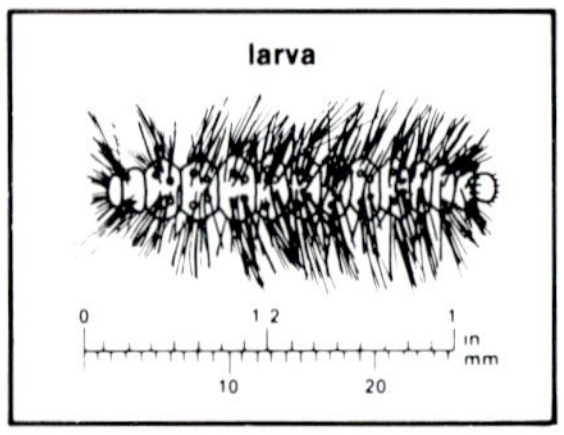

The larvae of the fall webworm spin silken tents high in the tree and feed inside on the leaves.

FALSE CHINCH BUG

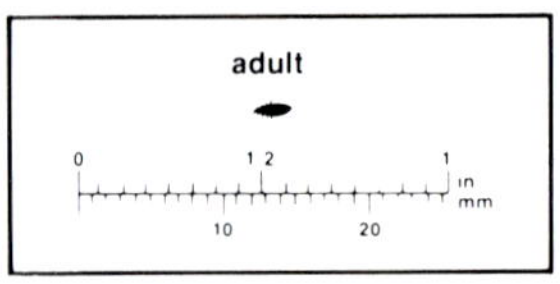

False chinch bug nymphs may feed on walnut leaves when their weed hosts dry up.

air applications are effective. The preferred material is *Bacillus thuringiensis*, a microbial pesticide that will not disrupt natural control. In orchards treated for the codling moth, the redhumped caterpillar is rarely a pest.

Fall Webworm
Hyphantria cunea

The fall webworm occasionally infests walnut orchards as well as black walnut, willow, poplar, mulberry and pecan trees. The larvae feed on leaves inside silken tents spun on branches, usually high in the tree. Heavy feeding can cause defoliation.

The fall webworm produces one generation a year; in some areas, it produces two generations. The first tents can be spotted in early summer. While feeding, the larvae enlarge their tents to include more leaves and entire shoots. Where two generations are produced, adult moths emerge in late summer. The eggs of this generation hatch in the fall. The webworm overwinters as pupae in cocoons on tree trunks or trash on the ground.

Monitor the webworms early in the summer, especially around the edges of the orchard. If you find a few colonies, prune them out and destroy them. Heavy populations rarely occur in orchards that are sprayed with insecticides in late spring or summer for other pests.

False Chinch Bug
Nysius raphanus

The southwestern false chinch bug is an occasional pest of young walnut orchards. Infestations usually develop where certain weeds, such as London rocket, shepherd's purse and common peppergrass, grow in late winter. The chinch bugs move into orchard trees as their weed hosts dry up in late spring. Because the preferred weed hosts reseed themselves in the same areas of the orchard, false chinch bugs tend to attack the same trees year after year.

False chinch bugs are "true" bugs with piercing, sucking mouthparts. They hibernate as adults and move in late winter to preferred weeds where they stay to lay eggs in early spring. The nymphs are dull gray or brownish red and collect in great numbers on the host plant. When nymphs move into trees, they may kill new foliage. This damage can occur within hours apparently because the nymphs inject a toxin while feeding. The leaves dry up and are covered with fecal spots. Damage may be substantial on trees one to three years old. Several generations are produced each year, but damage in orchards usually occurs only in spring.

Begin checking for the false chinch bug in weed stands in late February and early March before the walnut buds open. Monitor especially those areas where London rocket or other preferred weed hosts are growing. If you find adults and nymphs in large numbers, either in the orchard or in adjacent wild areas, disc these areas at least three weeks before the trees leaf out. This practice will deprive the bugs of food and shelter. If you detect the infestation only after the trees have leafed out, but before the bugs have moved into the walnut foliage, spray the weeds with an insecticide. If you find a large population of false chinch bugs after they have moved into the trees, treatment may be necessary. The last method of control is the least desirable because it disrupts populations of natural predators of other pests.

Minor Insect Pests

A number of other insects and mites occasionally inhabit walnut orchards, but they rarely cause significant damage. These include the European earwig, fire ant, flatheaded borer, filbertworm and blister mite.

European Earwig

The European earwig (*Forficula auricularia*) feeds on a variety of dead and living organisms, including insects, mites and growing shoots of plants. It occasionally damages expanding buds and leaves of newly budded and grafted walnut trees. You are not likely to see earwigs during the day because they spend the day in the soil at the base of the trees and feed at night. If you suspect that earwigs are causing damage, place a folded newspaper or burlap bag near the base of trees and check these traps the next morning. Earwigs rarely occur in numbers large enough to require treatment.

Southern Fire Ant

The southern fire ant (*Solenopsis xyloni*) may feed on walnuts after they have fallen or been shaken from the trees. They enter the nuts through the soft tissue at the stem end or through a codling moth injury. Rapid harvesting normally prevents damaging infestations.

Pacific Flatheaded Borer

The Pacific flatheaded borer (*Chrysobothris mali*) usually attacks only walnut trees that have been injured by sunburn, frost or other diseases. The larvae bore tunnels through the bark, cambium or wood. These tunnels sometimes girdle small branches or the trunk of young trees. Paint tree trunks with water-based white paint or whitewash to prevent sunburn and subsequent invasion by flatheaded borers.

Filbertworm

The filbertworm (*Melissopus latiferreanus*) is a rare pest in walnut orchards today. The adult moth is about the same size as the codling moth but has reddish brown or bronze forewings with a broad coppery band across the middle. The mature larva is white with a clear amber head. The larvae enter nuts after husk split and feed on the nutmeat.

Walnut Blister Mite

The walnut blister mite (*Eriophyes erinea*) is occasionally seen in walnut orchards. Its feeding on the lower surface of leaflets causes characteristic, blisterlike swellings on the upper surface of leaflets.

EUROPEAN EARWIG

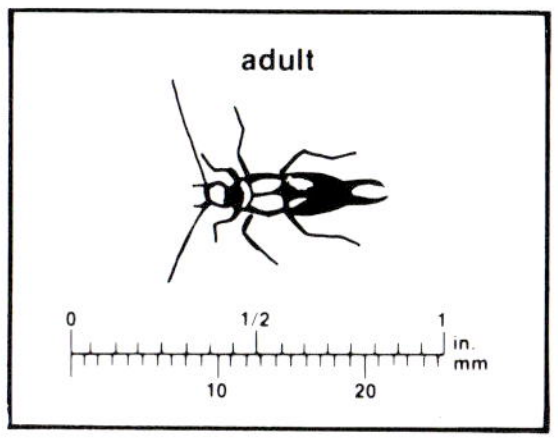

FIRE ANT

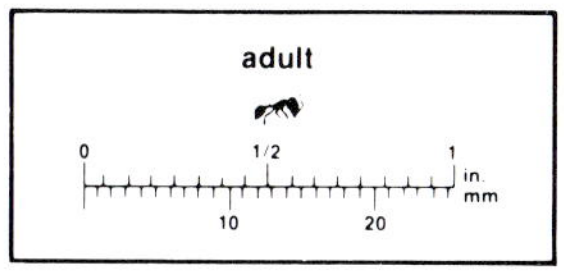

When walnuts lie on the ground, southern fire ants may enter them and chew tunnels into the nutmeat. Sometimes the sawdust-like chewings can be found in a nut.

PACIFIC FLATHEADED BORER

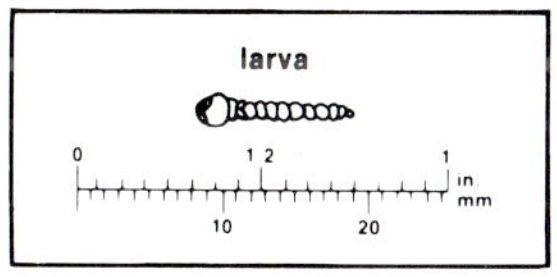

The larva of the flatheaded borer feeds inside twigs and branches.

Feeding of the walnut blister mite causes characteristic blisters on leaflets.

Blight infection on catkins starts in individual flowers, which turn black and die. During elongation, only the healthy part expands and the catkin becomes distorted.

Diseases

Walnut diseases are caused by a variety of microorganisms and environmental stresses. Blight, cankers and crown gall are bacterial diseases. Armillaria and Phytophthora root and crown rots are caused by fungi, and blackline is caused by a virus. Environmental stresses, such as insufficient or excessive water or minerals and adverse weather conditions, may contribute to the development of microbial diseases or produce disease symptoms of their own. These stress factors are discussed in a brief section at the end of this chapter. The cause of a disorder is often difficult to identify because trees may show nonspecific symptoms resulting from one or more pathogens, insects, nematodes or soil conditions. The severity of a disease depends on the virulence of the disease-causing organism (the pathogen), the susceptibility of the host and on environmental conditions.

The distribution of certain walnut diseases seems to have resulted more from changes in agricultural practices—bigger operations, more exchange of plant materials and equipment, and expansion to marginal soils—than from regional differences in climate or growing conditions. Some regional differences, however, do exist. Blackline was more frequent along the central coast but is now increasing in the Sacramento Valley and in the northern part of the San Joaquin Valley. Walnut blight is most severe in the humid coastal areas and in parts of the Sacramento Valley, where late spring rainfalls favor the spread of the disease. Deep bark canker is more common in the warm interior valley than in the coastal or mountainous areas.

Table 6 lists the major diseases and their sites of infection. Symptoms, however, may also occur in other parts of the tree, especially with trunk and root diseases, which interfere with water and nutrient transport, thereby causing leaves and shoots to yellow, droop or drop.

Table 6. Major Diseases of Walnuts and the Sites of Infection

Foliage and Nuts	Trunk and Limbs	Crown and Roots
blight	blackline	Phytophthora root and crown rot
	deep bark canker	Armillaria root and crown rot
	shallow bark canker	crown gall
	branch wilt	

Blackline and blight are the most important diseases. Blight causes direct economic loss by damaging flowers and growing nuts and providing breeding sites for the navel orangeworm. Blackline and other diseases that affect the crown, roots or trunk cause indirect damage to the crop. They limit nut yield or quality by reducing vigor and eventually kill the tree.

The pathogen, the host and the environment must all be considered in a disease management program. The pathogen, however, is difficult to control; thus, the emphasis usually lies on manipulating the host and its environment to prevent disease outbreaks or at least lessen the impact of a disease. Vigorous trees are able to resist infection or withstand a limited amount of pathogen activity. Prevention is especially important in disease management because many pathogens establish themselves slowly over the years. Once disease symptoms are apparent, few treatment options may remain.

When planning your management program, consider the following practices:

- choosing a rootstock or cultivar that is less susceptible to a particular disease or stress prevalent in the growing area (Tables 2, 3);
- pretreating the rootstock with a biological agent to protect against crown gall;
- planting walnuts on deep, well-drained soils; deep ploughing or back hoeing where layers or hard pans exist;
- adjusting the irrigation schedule to avoid saturating the soil; in poorly drained soils, splitting flood application or using sprinklers or drip irrigation; maintaining a ground cover to increase water penetration;
- applying copper sprays to prevent outbreaks of blight; avoiding irrigating during bloom; using low-angle sprinklers that don't wet leaves or nuts;
- avoiding surface spread of soil-borne pathogens by cleaning cultivation equipment, using sprinklers instead of flooding, and practicing no-till weed control in severely infested areas;
- practicing good weed control, especially around the base of trees where moisture favors the development of crown rot diseases, and during periods of water stress when competition from weeds can further weaken trees and make them more susceptible to certain diseases.

Walnut Blight

Xanthomonas campestris pv *juglandis*

Walnut blight, the major disease affecting leaves and nutlets, is caused by a bacterium that infects only walnuts. The disease occurs in all walnut-growing areas and may affect all cultivars, but it is most serious on early blooming cultivars, such as Payne and Ashley. Orchards that harbor blight may suffer heavy losses during years that are favorable to the development and spread of the pathogen.

Symptoms and Damage

All new growth, including buds, flowers, leaves and nutlets, is susceptible to walnut blight. Catkin infection begins in individual florets, which turn black, die and shrivel. As the infection spreads, part or all of the catkin blackens and becomes twisted. Female flowers are susceptible even before their stigmas are receptive for pollination. Infected flowers usually shrivel and drop from the tree.

Shoot infections may cause superficial lesions one to several inches long that can extend into the pith to form cankers. The lesions may girdle and kill the new twig.

Symptoms on leaves begin as small dark brown specks surrounded by a yellowish green halo. These infections kill small areas of the leaf during a period of rapid growth. Growing tissue folds around the lesions, giving the leaf a twisted appearance. Leaves infected after growth has been completed develop lesions but do not become distorted. As the infection spreads, larger areas of dark, dead tissue appear. Infected leaves remain in the tree, serving as sources of new infections that spread through the summer.

Early in the spring, young infected nutlets develop a dark lesion usually at the blossom end; most of these nutlets drop from the tree. As nuts mature, lesions can develop anywhere on the husk surface. These nuts remain on the tree. Husk lesions begin as small, water-soaked spots that later darken, enlarge, sink and often crack. When nuts become infected before the shells harden, the kernels may shrivel, making them unmarketable. Later infections will darken the kernels, lowering the quality of the harvest. Water stress and sunburn also cause shriveled or darkened kernels but produce husk symptoms different from blight; husks damaged by water stress are yellow to brown and often shriveled; sunburn lesions are brown and leathery and limited to the side exposed to the sun. Nuts infected by blight serve as a breeding site for the navel orangeworm, and they may also be more difficult to shake from the tree at harvest.

Seasonal Development

The blight bacterium survives the winter in twig cankers and in dormant buds and catkins. When growth resumes in the spring, the pathogen enters the new plant tissue through natural openings such as the stomata on catkins and new leaves. These primary infections produce more bacteria, which are spread to other sites in the tree, such as shoots, pistillate flowers, nuts and developing buds and catkins. This secondary spread may be repeated many times each season. The pathogen does

Shoots infected with blight show black lesions, which may girdle and kill new twigs.

Infection on the side of nuts starts as small, water-soaked spots.

Pistillate flowers usually become infected through the stigmas, which turn black and shrivel.

Infection sites on nuts later enlarge, blacken and often sink. Infected young leaflets become distorted as the growing tissue folds around the dark brown to black lesions.

not travel far inside the plant but causes local lesions. Windblown rain drops, sprinkler water or pollen carry the bacteria throughout the orchard. The severity of blight each season depends primarily on the presence of free moisture because the bacteria move and infect only when suspended in water. Blight can develop over a wide range of temperatures, but the spread of disease drops sharply when spring rains cease. Infections that occur later than May are associated with dew, sprinkler irrigation, or unseasonable rains.

Management Guidelines

The management program for blight focuses on preventing dispersal and development of the pathogen. Once infection has occurred, nothing can be done to stop bacterial growth at that site. Resistant cultivars are not available, but some are less susceptible to blight than others (Table 2). Generally, early blooming cultivars are more susceptible because wet weather during bloom favors bacterial growth and spread. In areas where late spring rains increase the chance for blight infections, mid- or late-blooming cultivars may be a wise choice. However, the most important management tactics involve adjustment in cultural practices in established orchards and the use of copper sprays.

Adjustment in cultural practices include:

- using low-angle sprinklers to avoid wetting of the foliage;
- avoiding irrigations during bloom; scheduling an irrigation during prebloom if winter rains were not adequate;
- removing low-lying limbs that are apt to catch the water from the sprinkler;
- avoiding crowding from double-planting of two cultivars by removing the cultivar that is more susceptible to blight;
- training young trees; heading the trees high to allow air movement under the trees and avoiding wetting low limbs;
- opening up trees during pruning to increase air movement through the tree.

Supplement these cultural practices with a regular program of preventive copper sprays in the spring. Copper, either as Bordeaux mixture or as a fixed copper spray, protects leaves and nuts by killing the bacteria before they infect; it cannot stop existing infections. Sprays must be frequent and provide good coverage to replenish the copper coating and to protect new growth.

A preventive spray of copper as the trees come into bearing and before blight occurs may prevent the buildup of inoculum in the orchard. Spraying alternate rows at each application date may be sufficient to reduce infection from blight at this stage.

In a mature orchard, it is a good practice to apply preventive copper sprays on a regular schedule, even if little or no blight damage occurred the previous year. The exact number of sprays necessary to achieve an acceptable level of control cannot be given and will vary with the orchard and the season. Apply the first spray at the earliest possible stage of pistillate bloom. In a low rainfall season, two to three applications during bloom may control blight, but in a wet season, sprays repeated at 7 to 10 day intervals through bloom or until rainfall stops are necessary. Although an intensive spray program may not eliminate blight in a wet year, even a limited number of sprays will reduce the blight damage.

In some areas, frequent spraying of copper, particularly Bordeaux, burns leaves. A summer oil emulsion added to Bordeaux can largely prevent phytotoxicity.

Blackline

Blackline was first a major problem in coastal areas but has since spread to the inland valleys and has become a threat to the walnut industry in many areas. The disease was long thought to be a problem of incompatibility between scion and rootstock. Recently, however, the causal agent has been identified as a strain of the cherry leaf roll virus.

Symptoms and Damage

The first symptoms of the disease are poor terminal growth, yellowing and drooping leaves and premature defoliation, particularly in the top branches. Later, diseased trees show dieback of terminal shoots and decline, often accompanied by profuse suckering of the rootstock.

Positive diagnosis of blackline requires careful examination of the union between scion and rootstock. Trees with blackline usually have small holes and cracks in the bark at the graft union. If you remove a piece of the bark, you can see a narrow black line or strip of dead tissue at the union. Blackline disease kills the tissues that transport nutrients and water between the rootstock and scion; thus leaves turn yellow and droop. In the early stages of the disease, the black line is not continuous around the union; eventually, it spreads around the graft union and completely girdles the tree, killing the scion two to six years later. The black line is usually exactly at the graft union. In addition to the black line, however, Paradox rootstock may develop a canker, which extends into the rootstock. Because of the extensive bark cankering, scions on Paradox decline faster than on black walnut rootstock. Blackline canker does not extend upward into the scion, as does the canker produced by *Phytophthora*.

Seasonal Development

The causal agent of blackline, the cherry leaf roll virus, can be spread in several ways. The virus may be introduced into an orchard through graft wood or pollen from trees infected with blackline. Within the orchard, the pathogen is probably transmitted by infected pollen because it does not spread from diseased to healthy trees until the pistillate flowers are receptive for pollination. In addition, blackline spreads more rapidly in mixed-cultivar orchards, where pollen shedding overlaps with pistillate bloom for a longer time. Spread of the pathogen through the soil or by nematodes is unlikely because black walnut and Paradox rootstocks are immune to the virus.

Trees may become infected at any age, but blackline is more common in trees 15 to 40 years old. After infection, trees decline over several years before dying.

Management Guidelines

At present, trees infected with blackline cannot be cured. No practical method is available to detect blackline before symptoms appear. A few management practices can help reduce the spread of the disease.

- Avoid interplanting with cultivars that develop pistillate bloom during the pollen-shedding period of already infected cultivars, especially in areas known to be infected with blackline.
- Select grafting material carefully from healthy trees to prevent introducing the pathogen into the orchard.
- Monitor the orchard for blackline symptoms in late summer or fall. Look for declining trees with sucker shoots at the base of trees. To confirm the presence of the disease, examine the graft union for the characteristic black line by making small cuts and removing the bark. Cut at 4-inch intervals because the black line may not be continuous.
- Remove the scion top of infected trees if only a few trees have blackline. Replace the whole tree or regraft the rootstock if it is healthy and vigorous. Choose a new scion cultivar whose pistillate flowers are not receptive during the major pollen-shedding period of infected cultivars.

Phytophthora Root and Crown Rot

Phytophthora spp.

This fungal disease, also known as crown rot, affects most deciduous fruit and nut trees. About eight different

The most characteristic symptom of blackline disease is the line of black, dead tissue at the graft union. At an early stage of the disease, the line may not be continuous and starts at one or several places around the graft union.

On Paradox rootstock, the blackline disease spreads into the rootstock as a canker. Note that the canker does not extend into the scion. On the right, you can see the black line at the graft union.

Trees infected with blackline usually show poor terminal growth, yellowing and drooping of leaves and suckering of the rootstock.

Phytophthora species are known to attack walnut trees. Some affect only the crown, some only the roots and others both the crown and the roots; more than one species may infect a single tree. Rootstocks vary in their susceptibility to the disease.

Symptoms and Damage

Trees affected with *Phytophthora* first show small leaves and sparse foliage and lack of terminal growth. The trees appear drought stricken early in the growing season and by midsummer the leaves turn yellow and droop, exposing the nuts to sunburn. Infected trees may decline for several years or die within the same growing season in which the foliage symptoms first appear. If diseased trees survive for several years, nut production gradually declines.

How soon the trees die after infection depends on the species of *Phytophthora*, and soil and water management and climatic conditions. The most common *Phytophthora* species in walnuts are *P. cactorum*, *P. citricola*, *P. megasperma* and three other unidentified *Phytophthora* species.

P. cactorum and *P. citricola* cause crown rot and trunk canker. A canker starts at one or several infection sites at the crown and spreads downward into the roots or upward into the trunk. It often moves up vertically in a narrow streak for several feet before spreading laterally, girdling the trunk. The canker extends through the bark and cambium and stains the wood. A black sap may ooze from the dark brown bark. To confirm the presence of crown rot, cut away the outer bark and look for a canker with water-soaked, concentric margins between healthy and infected tissue. *P. megasperma* and some of the unidentified *Phytophthora* species cause a slow decline due to root rot, which is mostly limited to feeder roots and small rootlets. The root symptoms are very subtle, especially in the early stages of an infection. Affected trees live for five years or longer before they finally die.

Another less common species, *P. cinnamomi*, infects the crown and the entire root system. Decaying roots are firm and brittle, whereas those destroyed by excess of water are soft and watery. Walnut trees with crown rot usually die sooner than those affected only with root rot. Trees infected by *P. cinnamomi*, *P. citricola* or *P. cactorum* die within one to five years.

Seasonal Development

Phytophthora can survive in the soil for many years. The fungus spreads and infects the trees during moist, cool weather in the spring and fall. The pathogen enters the tree either at the crown near the soil line, at the major roots or at the feeder roots, depending on the species. Those species infecting the roots cause root rot; those infecting the crown spread into the roots or the trunk where they form a canker. The canker develops rapidly under cool, moist conditions but spreads very slowly during warm, dry weather.

Management Guidelines

Eradicating *Phytophthora* from the orchard soil is extremely difficult. Fumigating the soil after you have removed a diseased tree is usually not completely successful. Careful orchard management can help minimize disease development and spread of the pathogen.

- Avoid planting in soil infested with *Phytophthora.*
- In soils infested with *Phytophthora,* use the Paradox rootstock. Paradox is significantly more resistant to *P. cactorum,* a major destructive species, and somewhat more resistant to other *Phytophthora* species than Northern California black walnut or English rootstock. Check with your farm advisor for help in isolating and identifying *Phytophthora* species.
- Good soil and water management is most important. Plant on soil with good surface and internal drainage. Use low volume sprinklers where possible. Where flood or furrow irrigation has to be used, split applications to avoid prolonged saturation (more than 12 hours) of the soil. Tensiometers are useful to find out when and how much water to apply. Drain water from areas where it tends to accumulate. Provide for good drainage of rain or irrigation water by planting the trees on a broad ridge 20 to 25 cm (8 to 10 inches) high.
- Be careful not to introduce the pathogen into uninfested soil through nursery material, irrigation water or farming equipment. Reduce traffic in the orchard, especially when the soil is damp and sticky. Clean farming equipment before moving from an infested area into a clean one.

Some *Phytophthora* species cause crown rot; if you remove the soil from the crown region, you can often see a black sap seeping from the infected bark.

To confirm the presence of Phytophthora crown rot, cut away the outer bark and look for a canker that has water-soaked, concentric margins and extends through the inner bark and cambium.

Armillaria Root and Crown Rot
Armillaria mellea

Armillaria root and crown rot, also known as oak root fungus disease, shoestring root rot and mushroom root rot, affects walnut trees and other native and cultivated plants. The severity of the disease depends on the walnut rootstock and the strain of *A. mellea.*

Symptoms and Damage

Symptoms may not develop until long after the disease is well established. The first symptoms of Armillaria root rot are poor growth or dieback of shoots, small yellowing leaves and premature leaf drop. The pathogen

A characteristic symptom of an infection with *Armillaria* is the creamy white, fan-shaped plaques of fungal mycelia underneath the bark (left). After rains in fall or spring, a cluster of mushrooms often appears at the base of infected trees (right).

invades the roots, crown and basal trunk, eventually girdling the crown region and destroying the entire root system.

The parasitic fungus invades the roots and crown with rhizomorphs. These root-like structures are black outside but white inside and resemble shoestrings. They grow on the surface of the bark and eventually penetrate the living tissue to form fan-shaped, creamy white plaques of mycelia between the bark and woody tissue at the crown region. Another wood-rotting fungus, a species of *Poria,* is often confused with *Armillaria. Poria* invades only through dead tissues, such as old crown galls, injuries, or diseased roots and does not produce rhizomorphs, fan-shaped plaques or mushrooms. *Armillaria* may form clusters of mushrooms at the base of infected trees a few days after heavy rain, especially in the fall. The mushroom caps range from 2 to 12 cm (1 to 5 inches) in diameter, are honey yellow to almost black and sometimes covered with brown scales; the gills underneath are light brown.

Seasonal Development

Armillaria develops most rapidly in moist, cool soil. The fungus is not spread by spores but rather by the rhizomorphs that grow along the roots from one tree to the next. Clusters of trees may be infected at one or several sites in the orchard.

The *Armillaria* fungus can survive for many years in dead or living roots of fruit and nut trees, oak, aspen and several fir and willow species. The disease often occurs along former stream beds and near creeks where the soil is moist and where dead roots or stumps harboring the fungus have been introduced.

Management Guidelines

Management of Armillaria root and crown rot relies primarily on preventing infection of new trees. Once infection is apparent, very little can be done to save a tree. You can reduce the chances of infection by carefully preparing planting sites for new orchards and by practicing good sanitation and early detection.

- Choose a less susceptible rootstock where *Armillaria* is a major problem. Northern California black walnut is usually resistant, although it is occasionally affected by *Armillaria.* Paradox is variable but on the average more susceptible than black walnut; English rootstock is highly susceptible.
- Avoid a planting site infested with *Armillaria.* Where this is not possible, remove and destroy all roots larger than 1.5 cm (½ inch). To prepare for fumigation, dry out the soil; this will improve penetration of the fumigant and also kill some of the fungus. Be aware that fumigants do not always kill all infected roots deep in the soil. See your farm advisor for advice on soil preparation, type of fumigants, application and precautions.
- Be careful not to introduce the pathogen into an established orchard in diseased root pieces, on equipment or in irrigation water.
- Check the orchard for symptoms of *Armillaria* in late summer when dead or declining trees are most obvious. Look for mycelial plaques under bark or mushrooms at the base of trees after a rain.
- Remove infected trees and adjacent, apparently healthy trees; once symptoms appear on one tree, the disease is likely to have already spread to the roots of the surrounding trees.

Crown Gall
Agrobacterium tumefaciens

Crown gall, a bacterial disease, can cause substantial damage to deciduous fruit trees and many other woody plants. Although crown gall can affect established orchard trees, the disease is most damaging to young trees. Black walnut is much more tolerant than Paradox, although neither rootstock is totally resistant to crown gall.

Symptoms and Damage

Trees infected with crown gall are easy to spot because they have distinctive galls at the base of their trunks or on their roots. These galls first appear as small outgrowths on the bark, generally at locations that have been injured from handhoeing or removal of sucker shoots. Young galls are smooth; as they age, they become rough and increase in size. Old galls are dark, brittle, and cracked. The galls consist of soft, undifferentiated tissue that lacks the rings of bark and wood found in healthy tissue or callus. If left unchecked, crown gall may progress around the crown, weakening and eventually girdling trees. Mature trees can tolerate a fair amount of gall development if infected in later years. Wood-rotting fungi, however, can enter through galls and further weaken trees.

Seasonal Development

When galls on trees infected with crown gall are broken or sloughed off, the pathogen enters the soil, where it survives for a short period. It can be spread through contaminated soil on equipment or plant material. The pathogen can only infect through wounds, such as caused by cultivation or pruning. Young trees in nurseries are particularly prone to infection because of the many potential injuries during rearing. The pathogen does not travel far inside the tree, and each gall is believed to be a separate infection site.

Management Guidelines

Because the disease organism infects the tree through injuries, crown gall can be largely prevented by practicing good sanitation, avoiding root and crown injury, and by using a protective agent. Soil fumigation does not eradicate the pathogen and is not recommended for control of crown gall in the orchard. The following practices are important.

- Handle young trees carefully and avoid excessive root injury.
- Dip young roots in a protective solution each time you dig or root-prune trees. A commercial preparation of *Agrobacterium radiobacter,* a biological agent, or a 1:20 dilution of household bleach can be used for this purpose. The bacterium provides a protective shield against the crown gall bacterium and is generally more effective than the bleach dip. Trees that are heeled-in and root-pruned may require dipping before each operation. Where only one treatment with *Agrobacterium* is economically feasible, use it before planting the trees in the field.
- Avoid injury to the crowns when cultivating, hoeing, or carrying out other orchard operations.
- Spray the cuts made when cutting off sucker shoots with a 1:20 dilution of household bleach. Suckers can also be burned off.

When established orchard trees are infected with crown gall, you can use a chemical that selectively kills the tumors. The treatment is most effective when the trees are not older than 3 to 4 years and when they are dormant. The procedure, however, is expensive and difficult to carry out. Many times it may be more economical to remove the tree and replant.

Branch Wilt
Hendersonula toruloidea

Branch wilt is a fungal disease of English walnut, almond, citrus, fig, European chestnut and poplar. It is widely distributed on walnut in the Central Valley.

Symptoms and Damage

The earliest symptom of branch wilt is the yellowing and withering of leaves on small, outermost branches during July and August. Next, all the leaves on the infected larger limbs suddenly wither, turn brown and dry up. These leaves remain attached to the twigs well after the other leaves have fallen, a trait that makes diseased trees easy to recognize in the fall. Portions of the thin outer layer of the bark peel away, exposing a black sooty mass of fungal spores. Diseased limbs have a gray to dark brown discoloration in the shape of a cylinder or partial cylinder extending into the center of the wood. If left unchecked for two or three years, many large limbs may be lost. Eventually the disease may extend into the trunk.

Seasonal Development

The fungal spores can invade bark only through splits, frost damage and, particularly, sunburn injury. Windblown or rain-splashed spores may enter such injuries throughout the year, but the disease only develops under warm temperatures. Warm nights and hot days favor fungal growth. The disease occurs most often on the southwest side of trees where sunburn injury is likely to occur. The fungus spreads within the branch to other limbs and may extend into the trunk.

Management Guidelines

Trees weakened by drought, nutritional deficiencies, crown or root rot or heavy mite populations are susceptible to sunburn and subsequent infection with branch wilt. The following management practices will help limit the incidence of branch wilt.

- Avoid sunburn by maintaining vigorous trees through adequate irrigation, fertilization and pest control.
- Check the orchard annually for branch wilt symptoms and immediately remove infected limbs; the best time to do this is in the fall after the leaves from healthy branches have dropped. Prune out diseased limbs and burn them. To make sure the disease does not spread, cut back into healthy, unstained wood, preferably for about a foot or more. Saws or other tools need not be disinfected; there is no evidence that pruning causes new infections.

CANKERS

The canker diseases, deep bark canker and shallow bark canker, are secondary problems brought about by other stresses that have weakened the trees, predisposing them to bacterial infection. Stress factors include shallow or compacted soil, drought, overirrigation, crowding or root and crown diseases; a combination of more than one may often be involved. Successful management of cankers involves identifying and then removing or reducing the stress factors rather than trying to cure the

The outgrowth at the base of this tree is a typical symptom of a crown gall infection.

A crown gall has been cut open to show the undifferentiated tissue inside.

An early symptom of branch wilt is the yellowing and withering of leaves on outermost branches in the summer. The leaves turn brown but do not drop.

canker symptoms. Neither canker disease, by itself, is known to kill trees.

Deep Bark Canker
Erwinia rubrifaciens

Deep bark or phloem canker was recognized as a disease distinct from shallow bark canker only in the mid-1960s when the causal bacterium was isolated. Deep bark canker is most common and most severe on the Hartley cultivar. The canker does not kill trees, but it may further debilitate trees already weakened by other disorders.

Symptoms and Damage

Trees infected with deep bark canker show symptoms of general tree decline, such as sparse or drooping foliage and poor nut yield. These general symptoms, however, are probably due more to stress factors preceding deep bark canker than to the bacterial infection itself.

The symptoms that characterize deep bark canker are the deep cracks running down scaffolds and trunks. A reddish brown to dark brown substance oozes from these cankers from late spring through early fall, giving them a "bleeding" appearance. Internally, dark brown to black streaks of varying width extend through the inner bark and may run for many feet up and down the scaffolds and trunks of affected trees. Because these streaks occur deep in the bark in the region of the phloem, the

The thin outer layer of the bark cracks and peels, exposing a sooty mass of spores of the branch wilt fungus.

Trees infected with deep bark canker have long, deep cracks, from which a reddish to dark brown sap oozes throughout the growing season.

Internally, the deep bark canker disease causes black streaks and typical small, dark pits in the inner bark.

The most characteristic symptom of shallow bark canker is brown to black blotches, which ooze a dark, tarlike substance.

Large superficial cankers may form just under the outer bark of trees infected with shallow bark canker.

disease was named phloem canker or deep bark canker to distinguish it from shallow bark canker, which is restricted to the outer layers of bark.

Another typical internal symptom of deep bark canker is the numerous small, round, dark-colored spots which extend into the wood beneath the cankered areas.

Seasonal Development

The pathogen survives the winter in cankers or dried exudate on the tree surface. The bacteria become active in late spring and begin to ooze from the cankers together with plant sap. The pathogen is spread by shaking equipment, possibly also by windblown rain and woodpeckers. It enters the tree most often through wounds caused by mechanical shakers. Because high temperatures favor the development of the disease, deep bark canker is more prevalent in the Central Valley than in the coastal areas.

Management Guidelines

To avoid predisposing trees to infection by deep bark canker, keep them healthy by practicing good water management, fertilization, pruning and pest control. These management practices will also reduce canker development in infected trees and keep them in production in most cases. Only trees growing on poor soils or under particularly adverse conditions may never fully recover. If you suspect deep bark canker in your orchard, the following guidelines can help you identify stress factors and reduce disease development.

- Dig a hole 2 to 3 meters (6 to 10 feet) deep in several areas of the orchard. Select sites that represent different soil types or growth habits. Study the soil profile for hardpans, compaction or sandy layers. Check the root depth and health of the root system.
- Install tensiometers in key locations in the orchard; they will aid in keeping track of soil moisture and adjusting irrigation schedules.
- Make sure the root zone is adequately moist, especially in the spring; a winter irrigation may be necessary in some areas or some years. At the same time, provide for adequate drainage.
- Remove trees in densely planted orchards when they start crowding.
- Avoid introducing the pathogen into the orchard; thoroughly scrub and drench the shaker pads in a 1:10 dilution of bleach before harvesting the trees.

Deep bark canker cannot be cured by any known chemical means. Cutting away the cankered areas both with and without applying copper or sodium hypochlorite has not proven successful. Attempts to halt the disease by injecting antibiotics have also failed.

Shallow Bark Canker
Erwinia nigrifluens

Shallow bark canker affects many commercial walnut cultivars, but the full range of susceptible hosts has not been determined. Because the damage is superficial, shallow bark canker is not considered a major disease of walnuts. It is not known how the pathogen infects walnuts or how it develops and spreads.

The most characteristic symptoms of shallow bark canker are brownish to black round spots or areas, usually several in a group, on the trunk or lower scaffolds. Newly infected areas have a margin of water-soaked bark and a central spot of black ooze that later dries, leaving a tarlike black spot. Just under the surface, dark brown areas of dying tissue are formed in the outer bark. These superficial cankers can be extensive, but they seem to cause little damage to the tree. Shallow bark canker rarely extends into the inner bark, as does deep bark canker.

As with deep bark canker, shallow bark canker often is severe in stressed trees. Improving tree vigor may help contain the disease. Cankers can be cut away, but this practice is not recommended because the disease is not serious enough to warrant possible wound damage and other infections.

Diseases Caused By Environmental Factors

Common disease symptoms of leaves and nuts may be caused not only by insects and pathogens but also by environmental stresses, such as adverse weather and soil conditions. Careful orchard management can alleviate these stresses and often prevent invasion by other pests.

Sunburn

If nuts are continuously exposed to the sun, the husk surface may show yellowish brown spots, which develop into dark brown, leathery lesions by mid- or late summer. Honeydew excreted by walnut aphids is an important factor contributing to sunburn on Payne and possibly other early cultivars because it kills the epidermal cells of the husk, causing the husk to turn black and more heat absorbent. Sunburned nuts have shriveled and darkened kernels. To minimize the sunburn damage to nuts, maintain a healthy foliage on trees and control aphids on Payne.

In addition to the nuts, the bark of young tree trunks and exposed limbs on older trees are also sensitive to sunburn. Sunburned bark is cracked and appears bleached. The damage may be confused with deep bark canker, but deep bark canker is not limited to areas exposed to the sun. Paint trunks of young trees with whitewash or a white, water-based paint to protect them against sunburn.

Molds

Molds may start to invade the nuts in midsummer if the husks have been damaged by sunburn, drought, or husk fly; damaged husks adhere to the shell and allow the mold to penetrate into the kernel, usually through the soft suture. The chance for mold infestation increases rapidly once the husks start to split. An early, light infestation is difficult to detect, but the organisms develop rapidly under warm, moist conditions. They form whitish gray mycelia on the surface of the kernel and often bluish green to grayish black spores, depending on the mold species. To prevent contamination by molds as much as possible, protect the husk from damage, and harvest and dry the nuts rapidly.

Windburn

If a strong, dry wind persists during the growth period of the foliage, the leaves in the most exposed areas develop small, dark lesions between the veins, twisting and folding the growing leaf. Many of these leaves drop. The mottling of the leaves may resemble zinc deficiency, but leaves deficient in zinc remain small. The Hartley cultivar seems to be most commonly affected, but the symptoms have been observed on other cultivars. Although this condition may appear severe, it does not seem to affect the crop. Nevertheless, check heavily windburned areas of the orchard for possible nutrient or water deficiencies or other stress problems that may contribute to the symptoms.

Frost and Freeze Damage

Frost and freeze damage is common in the interior valleys. Late spring frost can blacken and kill shots, budding leaves and flowers. Early fall frost may damage shoots and kernels, which darken and become rancid more rapidly.

The extent of frost damage depends on the temperature within the orchard and the physiological state of the trees. Soil condition, soil moisture and the presence of a ground cover may cause significant differences in temperature within neighboring orchards on cold nights. Firm soils that have recently been irrigated warm up readily during the day and radiate heat easily at night; orchards with dry or freshly disced soil or a ground cover will be cooler. Keep a ground cover mowed low and be prepared to irrigate during the frost season. Solid-set sprinklers are very convenient for such unscheduled irrigations. To protect against frost, some growers use heaters in their orchards.

The terminal growth, particularly of young trees, may be killed during winter freezes. In the spring, new shoots grow profusely below the dieback. Adequate moisture in fall and winter will lessen the impact of freezes. In a low rainfall year, a winter irrigation may be necessary.

Soil Problems

Soil problems can be minimized by planting walnut trees on deep, well-drained soils. Many problems arise because of waterlogging in the soil. Waterlogging leads to lack of oxygen, which causes decay of feeder roots. Despite excessive water, trees may look drought stricken because the uptake of water and nutrients is severely limited. Check the soil for layers of different soil texture; these may obstruct water drainage. If you use flooding on a heavy, layered soil, adjust your irrigation schedule or change to sprinkler or low volume irrigation techniques.

Nutritional Disorders

An excess or a deficiency of minerals in the soil is harmful to the tree because it interferes with the uptake of water and nutrients and with other physiological processes. Levels of minerals toxic to walnuts may be present in the soil or may build up because of poor quality irrigation water. Walnuts in particular have a very low tolerance for boron and chlorine (Table 4). Toxic levels of both minerals produce similar leaf symptoms: In mid- to late summer, tips and margins of leaflets turn brown; in severe cases, the burning extends between the veins.

Certain deficiencies also produce indicative leaf symptoms. Leaves deficient in zinc are finely mottled and remain small ("little leaf"). In a severe deficiency, these symptoms may appear as soon as the trees leaf out; in mild cases, symptoms appear from about May to August. Trees deficient in potassium have curled leaflets with gray margins. Iron deficiency is characterized by yellow to whitish, chlorotic leaflets on new growth. Symptoms of nitrogen deficiency may be confused with water stress: Early in the season, shoot growth is poor and leaves are pale and small. In early fall, the leaves turn yellow and drop prematurely. Other minerals, such as magnesium, phosphorus and copper, are rarely deficient.

Many symptoms appear either early or late in the year, so monitor the orchard throughout the season. Symptoms of marginal deficiency or toxicity or of combined nutritional disorders are very difficult to diagnose and should be confirmed by a leaf analysis performed by a professional lab.

Husks exposed to the sun develop yellowish brown to dark brown, leathery lesions during the summer.

S. SIBBETT

Frost can kill foliage and, in severe cases, entire young trees.

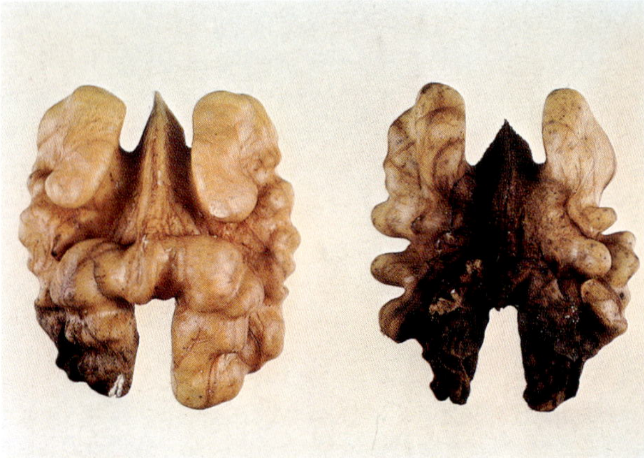

Sunburn can lead to darkened and shriveled kernels.

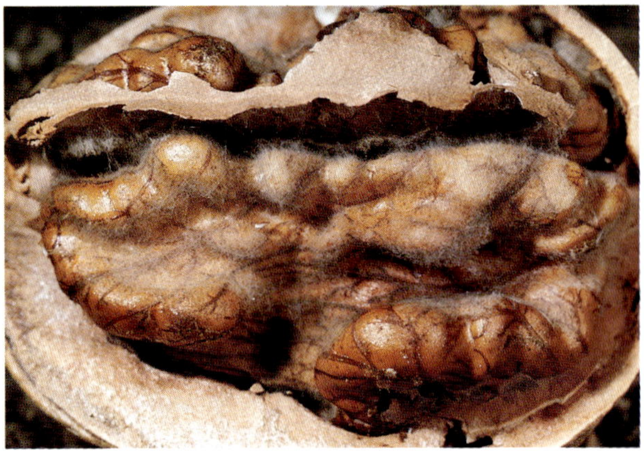

Molds often invade nuts that are damaged, but they may also enter sound nuts after husk split. Under warm, moist conditions, molds develop rapidly and form whitish gray mycelia on the kernels.

Strong, dry winds during the growth period of the foliage can cause brown lesions between the veins, often twisting the expanding leaflet.

K. URIU

Excess boron or chlorine results in brown leaf tips and margins by mid- to late summer. Both toxicities may occur together, as in this leaf sample.

K. URIU

If zinc is deficient, leaves are finely mottled and remain small ("little leaf").

K. URIU

Leaves deficient in potassium have curled, gray margins in mid- to late summer.

K. URIU

Iron deficiency is characterized by light yellow, chlorotic leaves.

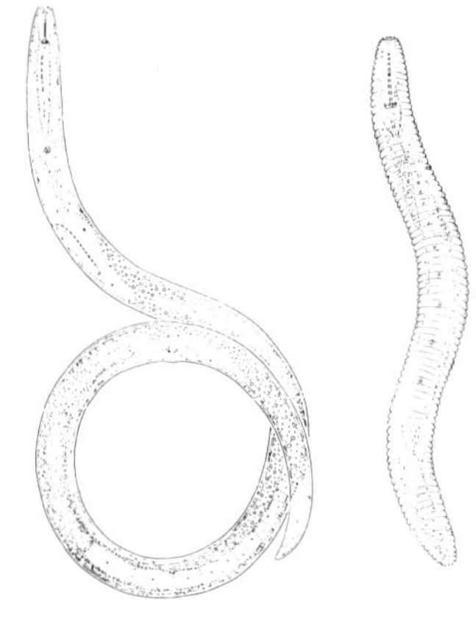

FIGURE 34. The two most common nematode species damaging walnut roots are the root-lesion nematode (left) and the ring nematode (right). Life sizes are 0.6 mm and 0.5 mm, respectively.

FIGURE 35. The root-lesion nematode causes black lesions on larger roots. Scrape the bark off the root with a knife to confirm that the lesions extend into the secondary phloem (top).

Nematodes

The extent of the damage that nematodes cause to walnut trees depends largely on the density of the nematode population and the age of the trees. Young trees planted in lightly infested soil may grow for many years without apparent damage. If a heavily infested orchard site is replanted, the roots of the young trees will soon be heavily parasitized, resulting in little or no growth.

Mature walnut trees can tolerate a fairly high number of these small root parasites. Adverse effects become apparent when the nematodes reach numbers that damage and kill roots faster than they can be regenerated. At this point, trees begin to decline. A deficiency or excess of water or nutrients, mechanical root damage and other root diseases can aggravate the impact of nematodes.

The above-ground symptoms of a nematode infestation are difficult to diagnose. Lack of vigor, small leaves, dieback of twigs and reduction of yield can also be caused by adverse soil conditions or by pathogens. The principal symptom that indicates a nematode infestation is the sparse root system under the tree, particularly the lack of small feeder roots. Extraction and identification of nematodes from the soil and roots is necessary to confirm nematodes as the limiting factor in tree growth.

Nematode Species in Walnuts

The two species of nematodes that most often parasitize the roots of walnut trees are the root-lesion nematode (*Pratylenchus vulnus*) and the ring nematode (*Macroposthonia xenoplax*) (Figure 34). Root-knot nematodes (*Meloidogyne* spp.) can cause swellings or galls on roots and are most important on English rootstock. Since this rootstock is not widely used, the root-knot nematode is currently not a concern to most walnut growers.

The root-lesion nematode is the most important species. It is found in almost all old walnut plantings and in any soil type. Root-lesion nematodes are endoparasites; they feed, migrate, and reproduce inside roots. This activity causes black lesions, which are very apparent when the roots are scraped with a knife (Figure 35). In the advanced stage, feeder roots are destroyed, severely limiting the tree's ability to take up water and nutrients. The decline in vigor can predispose the tree to sunburn and bark cankers and cannot be corrected by any management practices. In fact, adding more water or fertilizer may further damage weakened roots and invite other problems, such as root and crown rots.

The ring nematode is a migratory ectoparasite; it lives and reproduces in the soil while feeding on root tips and hairs with its long, spearlike mouthparts. The ring nematode can build up to large numbers in coarse, sandy-loam soils and causes mainly stunted feeder roots, which often proliferate and form dense mats.

Management Guidelines

Prevention. Before planting a walnut orchard, consider the cropping history of the site. If the site has recently been cleared of trees or vines, the soil is likely to be heavily infested with nematodes and is a poor site for a new walnut planting. A grower adamant about planting on such a site can reduce the nematode population by fallowing or growing an annual crop and then fumigating and planting a tolerant rootstock. The rotation allows the decay of woody roots, which harbor nematodes and protect them from fumigation. For most effective fumigation, carefully observe label instructions concerning land preparation, soil type, temperature, wind, residual crop roots, dosages and waiting periods between treatment and planting. Fumigation will not kill all nematodes at the application rates that are economically feasible. If the soil was heavily infested before fumigation, the nematodes will build up again to large numbers in a few years.

Rootstocks vary in their susceptibility to nematode damage. The Paradox hybrid is most tolerant to infestation by the root-lesion nematode, but this tolerance is not high enough to sustain vigor in highly infested soils. English and black walnut are easily affected by the root-lesion nematode; English rootstock is also susceptible to root-knot nematodes.

Good sanitation practices are important in preventing an infestation with nematodes. Buy your trees from a nursery that sells material free of nematodes. Clean equipment thoroughly when moving into another orchard. Do not irrigate with runoff water from areas infested with nematodes.

Little is known about how to use cultivation and irrigation practices to prevent a buildup of harmful nematode populations in an orchard. In general, any practice that favors root growth also favors an increase in nematodes. Environmental conditions and biological antagonists, such as nematode-trapping fungi, predaceous nematodes and insects, normally play a role in reducing populations of harmful nematodes. No methods, however, are known for increasing this role sufficiently to keep nematode populations below damaging levels.

Monitoring and Control. Nematodes are so small that you cannot detect most of them with a hand lens. If you suspect damage by nematodes or want to know the status of a site before planting, collect soil samples for professional analysis. For listing of professional laboratories, see UC Agricultural Science Publication 3024.

To collect samples before planting, visually divide the orchard into sampling blocks representing differences in soil texture, drainage pattern or cropping history. In an established orchard, collect samples in areas where symptoms appear and compare them with samples from an adjacent area where trees look healthy.

Sampling can be done at any time of the year. Sample when the soil is moist. Take samples with an auger, sampling tube or shovel. From each sampling block, collect 10 to 20 cores or subsamples at a depth of 15 cm to 1 meter (6 inches to 3 feet). In an established orchard, take subsamples at the edges of the canopy and include some feeder roots. Combine the subsamples, mix thoroughly and take one quart as a representative sample for this block. Pour the soil and roots into a durable plastic bag or other moisture-proof container. Keep it in a cool place (10 ° to 15 °C or 50 ° to 60 °F). Attach a label providing name and address, location of the orchard, soil texture, cropping history and notable symptoms; this information is critical for a meaningful analysis. Send or deliver the sample to the lab as soon as possible. Ship in a cardboard box insulated with newspaper or in a styrofoam ice chest.

There are no known methods for controlling nematodes once you detect their damage in a walnut orchard. The degree of damage and the speed with which the infestation spreads depends on how many nematodes were present at the time of planting and on how they were introduced into the orchard. If introduced on seedling roots, nematodes may increase and be destructive in a few years. If introduced in contaminated soil, equipment or runoff water, the nematodes need to establish themselves and may take 10 to 15 years to reach damaging levels.

Remove weeds growing around tree trunks; they hold moisture and increase the chance for crown diseases.

Mowing or flailing several times during the year is one way to manage weeds in orchards.

C. ELMORE

Strip weed control in orchards involves using herbicides along tree rows and mowing or discing between the tree rows.

Weeds

Plants growing on the orchard floor become weeds when they compete with walnut trees for water, nutrients and sunlight, contribute to pest problems or interfere with cultural operations and harvest. Competition is particularly strong in young orchards. Young trees stressed for water and nutrients do not grow as fast as they could and may be less tolerant of damage from insects and disease. Weeds around tree trunks also hold moisture and increase the chance for crown rot infections. As the trees grow older, their canopies shade most of the orchard floor, greatly reducing the growth of summer annuals. Weeds have to be controlled in bearing orchards to allow efficient irrigation and mechanical harvest and to reduce stress in drought periods. For beneficial effects of maintaining plant growth on the orchard floor see page 16.

The weed species found in walnut orchards can be grouped as annuals and herbaceous perennials (Figures 36–38). Annual plants germinate, grow, flower, and produce seeds in one season. Winter annuals germinate in the fall, grow during the winter, and flower and go to seed in the spring before dying in early summer. Common winter annuals found in orchards are annual bluegrass, chickweed, common groundsel, mustard, nettle, pineapple weed, wild barley and wild oats. Summer annuals germinate in the spring or early summer and flower and produce seed in the fall before dying in the winter. Major species include barnyardgrass, crabgrass, knotweed, common lambsquarters, pigweeds, puncturevine, purslane, sowthistle and sprangletop. In California's mild climate, certain annuals may overwinter and behave as short-lived perennials.

Winter annuals present the least problems because they are growing during the winter when ample moisture for tree growth is generally available. They can be managed by cultivating or mowing in early spring or by using herbicides. Summer annuals and perennials, however, require more intensive management because they compete with trees for water during the dry summer months and may interfere with irrigation and harvesting operations.

Perennials, which can live three years or longer, may be herbaceous or woody plants. The herbaceous plants die back during the winter but regrow during the spring or early summer. Commonly found herbaceous perennials in orchards are bermudagrass, curly dock, dallisgrass, field bindweed, and johnsongrass.

Perennials are difficult to control because they usually have extensive stolons, rhizomes, bulbs, tubers or tap roots, which can regrow new plants after mowing or cultivation. Control measures depend on the type of plant (grass or broadleaf) and its growth stage. Generally, it is easiest to control perennials before they have formed rhizomes. If established perennials are present, they should be treated before they produce seeds. A control program usually involves a postemergence herbicide application and follow-up cultivation or preemergence treatment for the seedlings.

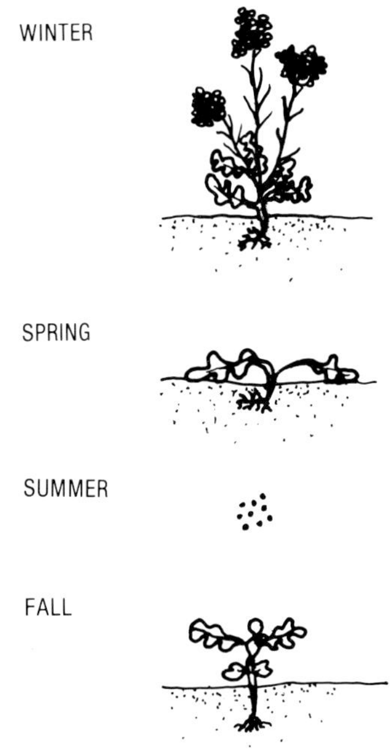

FIGURE 36. **Winter annuals germinate in the fall, mature in the winter and die in early summer. The seeds remain dormant until the fall.**

Management Guidelines

Prevention.

Careful management practices and sanitation in and around the orchard can help limit weed infestations. To avoid the spreading of weeds, make sure irrigation ditches are free of weeds. Clean cultivating and harvesting equipment. Drain areas with high moisture or standing water because they provide ideal conditions for dallisgrass and curly dock. By letting the top two to four inches of soil dry completely between irrigations, you can discourage weed growth, especially the establishment of seedlings.

Biological control of weeds in orchards so far is very limited. Puncturevine stem and seed weevils have been established, and they help reduce the number of puncturevines in orchards. The larvae of the purslane sawfly feed inside purslane leaves, but their effect on the plant is not known.

Control Methods

Weeds in orchards can be managed by total cultivation, total mowing, by strip weed control with herbicides, or by total reliance on herbicides. The advantages and disadvantages of each method are discussed below.

Total Cultivation. Total or clean cultivation (tillage) keeps the orchard free of weed growth during most of the year. Disc or harrow the orchard floor once during the late winter or early spring and four or more times during the summer, depending on the number of irrigations and the amount of weed growth. You will probably have to weed around the trunks by hand several times a year, especially if your orchard is young. To prepare a weed-free orchard for harvest, disc and roll or float the orchard floor.

Weeds allowed to grow during the winter will prevent soil erosion and improve soil structure. However, where spring or fall frosts are a problem, clean cultivation provides a warmer orchard. If water is limited during the summer, shallow cultivations will help conserve water.

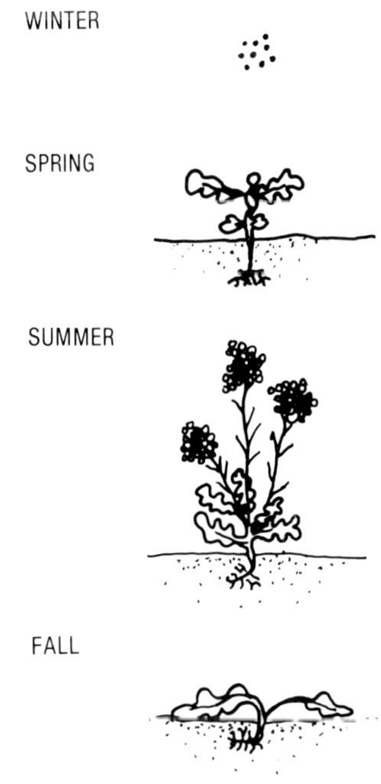

FIGURE 37. **Summer annuals germinate in the spring, mature in the summer and die in the fall. The seeds remain dormant until the spring.**

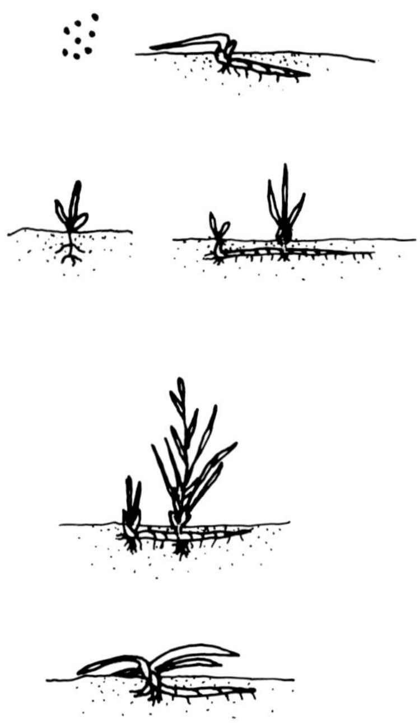

FIGURE 38. **Herbaceous perennials grow new plants from seeds or vegetative parts, such as rhizomes, bulbs, tubers or rootstocks, in the spring. They mature in the summer and die in the fall; seeds and underground parts overwinter.**

Total cultivation, however, requires discing in both directions with heavy equipment that tends to compact the soil, decreasing water penetration. The discing cuts the feeder roots in the top six inches of soil; thus, the tree cannot use the rich supply of nutrients, air and water in this area. The discing may also injure the lower trunk, making the trees susceptible to crown diseases. Cultivation also generates dust which often increases mite problems.

Total Mowing. The total mowing regime utilizes frequent mowing to manage orchard floor vegetation, which can be a cover crop or a sod culture. A cover crop is a temporary cover provided by annual weeds or, in some cases, by a sown grass or legume. A sod culture consists of a permanent ground cover provided by volunteer annual or perennial plants. Mowing or flailing should be done each time the plants reach 12 to 20 cm (6 to 8 inches), that is, before seed heads form on tall growing weeds. This will allow desirable, low growing winter annuals, such as redmaids, filaree, annual bluegrass and shepherdspurse, to reseed themselves. Mowing about once a month during spring and summer is usually sufficient, but more frequent cutting may be necessary in some areas, especially during the establishment of an orchard. To avoid injuring the tree trunks with the mowing equipment, weed around the trunks by hand or with herbicides. A few weeks before harvest, the ground cover is mowed very close to the ground to create a smooth surface for mechanical harvest.

Total mowing offers certain advantages. This system uses lighter equipment than the total cultivation system, reducing soil compaction and making the orchard accessible during wet conditions. The roots of the cover plants help maintain good water penetration. Tree roots can grow closer to the surface to make use of water, air and nutrients in the top soil. A ground cover also reduces dust.

On the other hand, there are some drawbacks: If mowing is not done in time, the mowed material will not completely decay, leaving trash. The cooling effect of the ground cover can be advantageous in areas with hot summers but a disadvantage in areas with frost hazards. Maintaining a ground cover also requires more water and nutrients than a clean-cultivated orchard. Under mowing, perennials tend to become the dominant weed population because they grow longer and compete better for nutrients and water than annuals.

Strip Weed Control. Keeping weeds away from the trunks is difficult to do in total mowing or total cultivation programs without injuring trees. Using strip weed control, however, solves this problem: apply herbicides along tree rows to keep a 3- to 6-foot wide strip weed free; between the rows, mow or disc weeds.

	PREEMERGENCE								POSTEMERGENCE			
	simazine	diuron	norflurazon	napropamide	oryzalin	trifluralin	dichlobenil	EPTC	paraquat	glyphosate	dinoseb	MSMA
ANNUAL GRASSES												
annual bluegrass	partially controlled	controlled	controlled	controlled	controlled	controlled	partially controlled	controlled	controlled	controlled	not controlled	controlled
barnyardgrass	partially controlled	controlled	partially controlled	partially controlled	controlled	controlled	controlled	controlled	partially controlled	controlled	not controlled	partially controlled
crabgrass	not controlled	controlled	controlled	controlled	controlled	controlled	not controlled	controlled	controlled	controlled	not controlled	controlled
johnsongrass (seedling)	not controlled	controlled	controlled	controlled	controlled	controlled	not controlled	controlled	controlled	controlled	not controlled	controlled
wild barley	partially controlled	controlled	controlled	controlled	controlled	controlled	controlled	controlled	controlled	controlled	not controlled	ND
wild oats	not controlled	partially controlled	controlled	controlled	partially controlled	partially controlled	partially controlled	partially controlled	controlled	controlled	not controlled	partially controlled
ANNUAL BROADLEAVES												
cheeseweed (mallow)	partially controlled	controlled	ND	partially controlled	partially controlled	not controlled	controlled	partially controlled	not controlled	partially controlled	partially controlled	not controlled
clovers	controlled	controlled	ND	partially controlled	partially controlled	not controlled	controlled	not controlled	partially controlled	partially controlled	controlled	not controlled
common groundsel	controlled	not controlled	ND	controlled	not controlled	not controlled	controlled	controlled	controlled	controlled	partially controlled	not controlled
common lambsquarters	controlled	controlled	controlled	controlled	controlled	controlled	controlled	controlled	partially controlled	controlled	controlled	not controlled
fiddleneck	controlled	controlled	ND	controlled	controlled	controlled	controlled	ND	controlled	controlled	controlled	not controlled
filaree	controlled	controlled	ND	controlled	partially controlled	partially controlled	partially controlled	partially controlled	partially controlled	partially controlled	partially controlled	not controlled
flax-leaved fleabane	controlled	controlled	controlled	not controlled	not controlled	not controlled	controlled	controlled	controlled	controlled	partially controlled	not controlled
horseweed	controlled	controlled	controlled	not controlled	not controlled	not controlled	controlled	controlled	controlled	controlled	partially controlled	not controlled
knotweed	controlled	controlled	ND	controlled	controlled	controlled	controlled	ND	partially controlled	partially controlled	partially controlled	not controlled
mustards	controlled	controlled	controlled	not controlled	not controlled	not controlled	controlled	partially controlled	controlled	controlled	controlled	not controlled
pigweeds	controlled	controlled	not controlled	controlled	controlled	controlled	controlled	controlled	controlled	controlled	controlled	not controlled
puncturevine	not controlled	not controlled	controlled	partially controlled	partially controlled	partially controlled	partially controlled	not controlled	controlled	ND	not controlled	controlled
purslane	controlled	controlled	controlled	controlled	controlled	controlled	controlled	controlled	controlled	controlled	controlled	not controlled
PERENNIALS												
bermudagrass	not controlled	not controlled	partially controlled	not controlled	partially controlled	partially controlled	not controlled	not controlled	not controlled	controlled	not controlled	not controlled
dallisgrass	not controlled	not controlled	not controlled	not controlled	not controlled	not controlled	not controlled	not controlled	not controlled	controlled	not controlled	partially controlled
johnsongrass	not controlled	not controlled	ND	not controlled	partially controlled	partially controlled	not controlled	not controlled	not controlled	controlled	not controlled	controlled
field bindweed	not controlled	not controlled	ND	not controlled	partially controlled	partially controlled	partially controlled	not controlled	not controlled	partially controlled	not controlled	not controlled
nutsedge	not controlled	not controlled	ND	partially controlled	not controlled	not controlled	controlled	partially controlled	not controlled	partially controlled	not controlled	partially controlled

not controlled · partially controlled · controlled · ND = no data

FIGURE 39. Weed control provided by herbicides currently registered for use in walnuts. This information is derived from field experience and is tentative; changes may be necessary as new information becomes available.

When treating weeds along tree rows, apply long-lasting preemergence herbicides, which act on the seedlings as they germinate, in the fall or early winter. If you use sprinkler irrigation, you can split the rate with two treatments: one in the fall or early winter and the other in summer. If established weeds need control, add a postemergence herbicide.

You can use this system in both young and established orchards and with drip, sprinkler or flood irrigation with permanent berms. Strip weed control reduces the amount of herbicide needed and saves time and energy because it requires only one-way mowing or discing. It also avoids injuring trees and spreading diseases and offers all the advantages listed under total mowing. When practicing strip weed control, watch for perennials, especially in untreated areas. Where perennials appear, consider a spot treatment. Once perennials have spread, they are very difficult to control and may move into the strip.

Total Reliance on Herbicides. The orchard floor can be kept free of weed growth if you spray the entire ground with herbicides. Although this practice is convenient, it may be more costly and in some soils may reduce water penetration.

Use of Herbicides

Herbicides are applied differently according to their mode of action. Preemergence (soil-applied, residual) herbicides are applied to the soil surface and must be incorporated into the soil mechanically, by irrigation or with rainfall. They kill weeds as they germinate. Some materials may suppress root growth of trees if mixed into the root zone. Most, however, are applied to the surface and do not reach the roots. Because these herbicides can persist in the soil from a few months to a year or more, discontinue using them one or two years before removing an orchard. Where it is necessary to replace a tree, make sure to backfill with untreated soil around the roots of the young tree. Preemergence herbicides degrade faster if the soil is kept moist, such as under low volume irrigation.

Postemergence (foliar-applied) herbicides are used on established weeds. They act either by contact or by translocation throughout the plant. Contact herbicides, such as weed oils, kill those parts of the plant that are actually sprayed; good coverage is therefore essential. Contact herbicides are most effective with the least amount of material on weed seedlings or young weeds. Because of their different mode of action, translocated or systemic herbicides work best on vigorously growing weeds. Postemergence herbicides, especially weed oils, will injure the tree if applied to the foliage, shoots or young bark.

Often combinations of herbicides provide good control while saving on application time or material. If you need to control established weeds at the time of preemergence treatment, add a postemergence herbicide. You can often use lower rates of weed oil when combined with low levels of a synthetic postemergence herbicide. Figure 39 shows the susceptibility of major weeds to herbicides currently registered for use in walnuts. For details of registration status and use of individual herbicides, see the UC Agricultural Science Publication on weed control in walnuts.

Monitoring

To choose the correct herbicide or combination of herbicides and the right time to apply them, monitor your orchard for weeds three times a year, in February, May and November. In November (after the first rains), look for seedlings of winter annuals to determine what type of preemergence treatment is necessary. In February, you will be able to record the full spectrum of grown winter weeds. If a few winter weeds were not controlled by the preemergence treatment, note them for possible changes in next season's treatment. If weed growth is heavy, you may consider postemergence treatment. By May, summer annuals will have germinated. Monitoring at this time will show if preemergence treatment for summer annuals was successful and what species of perennials are present. Keep records of your observations; a sample format is given in Figure 40. The records will give a good overview of the weed species present and allow you to evaluate past practices and to plan future actions.

Identify the weeds, preferably in the seedling stage, and keep records of predominant species in the orchard. The following descriptions and photos should help in identifying the major species found in walnut orchards. A few necessary technical terms are illustrated in Figure 41 and explained in the glossary. Broadleaf weeds have two seed leaves, often of characteristic shape, color and texture. The seed leaves and usually also the first true leaves differ from later leaves. Grasses have one seed leaf. The collar region is usually the most distinguishing feature of true leaves in grasses; perennial grasses also have characteristic rhizomes and often stolons. In most instances, you have to use a combination of characteristics to identify a species. Pick several young plants, for the characteristics may be more or less pronounced in individual specimens, depending on their growing conditions. If you encounter weed species not covered in the chapter, check with UC Agricultural Sciences Publication 4030, *Growers' Weed Identification Handbook*, your farm advisor or your agricultural commissioner.

ORCHARD LOCATION ____________________ **CONTROL METHODS** ____________________

CONTROL DATES ____________________

COMMENTS ____

	NOV ______		FEB ______		MAY ______	
	% of total weeds		% of total weeds		% of total weeds	
	treated	untreated	treated	untreated	treated	untreated
ANNUAL GRASSES						
annual bluegrass						
barnyardgrass						
crabgrass						
sprangletop						
wild barley						
wild oats						
______						
______						
______						
ANNUAL BROADLEAVES						
cheeseweed (mallow)						
clovers						
groundsel						
filaree						
fiddleneck						
knotweed						
lambsquarters						
mustards						
pigweeds						
puncturevine						
purslane						
______						
______						
______						
______						
______						
PERENNIALS						
bermudagrass						
dallisgrass						
johnsongrass						
curly dock						
field bindweed						
nutsedge						
______						
______						

FIGURE 40. Keep records of the weed species present in your orchard on a form such as this one. Monitor three times a year and note the percentage of each weed in relation to the total weed population. If you have treated and untreated areas, such as in strip weed control, record the weed species in each area separately.

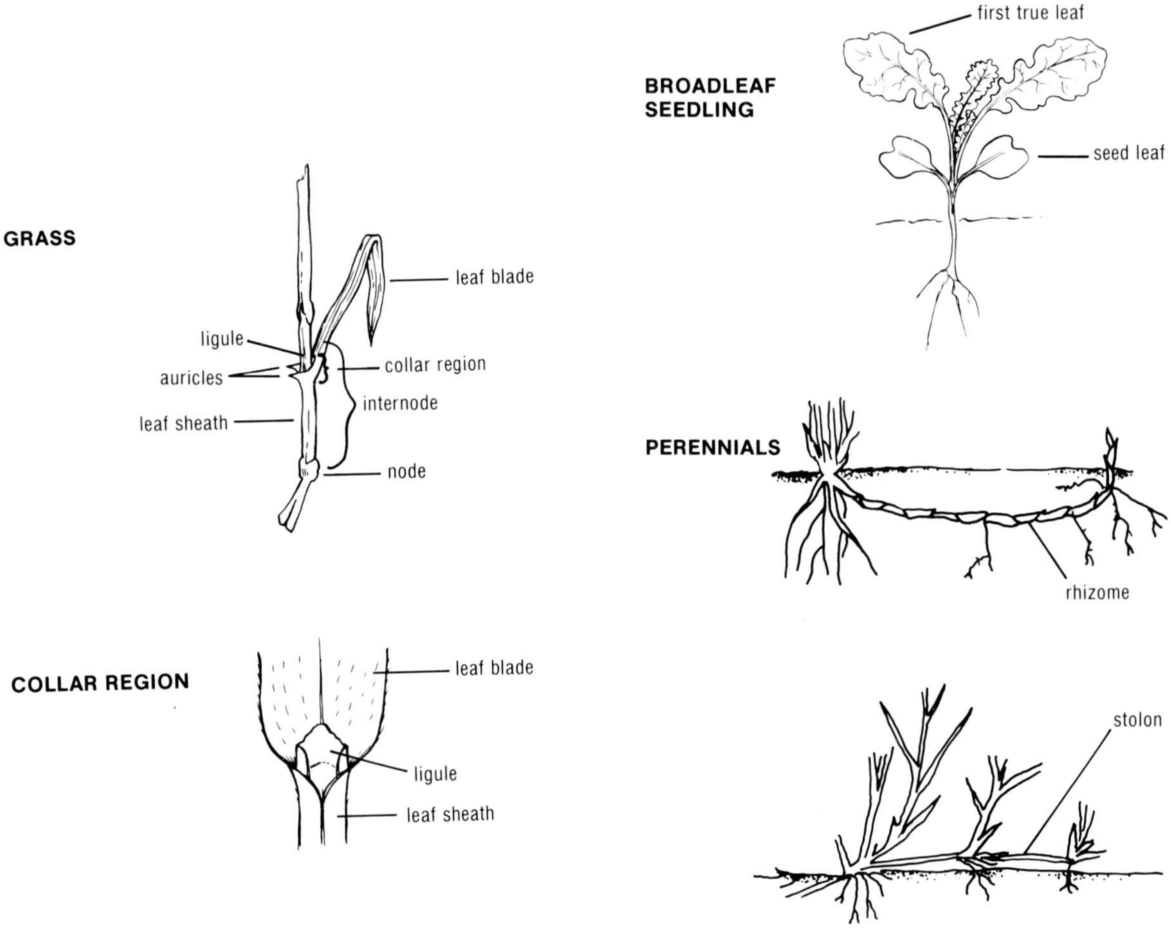

FIGURE 41. Major identifying characteristics of weeds.

Winter Annuals, Grasses

Annual Bluegrass *Poa annua*

Annual bluegrass provides an excellent ground cover because it is low growing and dies early in the summer when the top soil dries out. It can easily be distinguished from other grasses by its typical leaf tip and blade.

A. The mature plant grows as dense, low-spreading tufts, 8 to 30 cm (3 to 12 inches) tall, and often roots at the lower nodes.

B. The most distinguishing feature of the mature plant and the seedling is the blunt leaf tip shaped like the bow of a boat. The leaf blade is often crinkled at midsection.

A

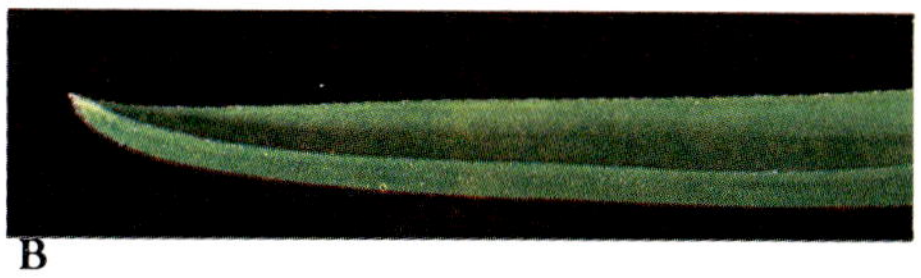

B

Wild Oats *Avena fatua*

Wild oats is troublesome in orchards because its high growth around tree trunks holds moisture, making the crown susceptible to diseases. It dries up and dies in midsummer. The seedlings of wild oats and wild barley look very similar. Wild oats, however, retains for a long time the seed from which it is growing. Mature plants of the two species are easily distinguished by their different flower heads.

C. The collar region has a tall, pointed ligule with toothed margins and no auricles (right); also shown are wheat (middle) and wild barley (left).

D. The mature plant is sturdy, 30 to 120 cm (1 to 4 feet) tall, with several stems arising from the base.

C

Wild Barley *Hordeum leporinum*

Wild barley is less bothersome than wild oats because it is much shorter and dries up earlier in the year. Wild barley has characteristic narrow, long auricles, which clasp the stem (D, far left); the auricles, however, may not be very pronounced in the young plant.

E. The mature plant grows as short, spreading tufts, 15 to 60 cm (6 to 24 inches) high. The flower head is a dense, bristly spike with erect awns; at maturity, the head breaks apart into several pieces.

D

E

A

B

C

D

E

F

G

Winter Annuals, Broadleaf

Common Groundsel *Senecio vulgaris*

Common groundsel has increased in abundance in vineyards and orchards because of its tolerance to some herbicides. It can be distinguished from other similar looking members of the thistle family by the seed leaves and first true leaves.

A. The bracts at the base of the flower clusters have typical black tips. The mature plant is erect, branched, 50 to 60 cm (20 to 24 inches) tall.

B. The seed leaves are smooth and elongated oval in shape. The first true leaves have shallow teeth arranged in steplike fashion; later leaves are more deeply lobed. The seedling has a fairly strong tap root.

Little Mallow *Malva parviflora*

In California's mild climate, little mallow or cheeseweed usually overwinters and is therefore sometimes considered a biennial. It thrives on nutrient-rich and sandy loam soils. Because of its tolerance to many herbicides, it often becomes the predominant ground cover in orchards. Several mallow species occur in California.

C. The flowers are held in clusters in the leaf axils; the fruit resembles a miniature wheel of cheese. The mature plant forms dense bushes that trail along the ground or grow upright, 25 to 125 cm (10 to 50 inches) tall.

D. The seed leaves are distinctly heart shaped and tinged red. The true leaves are roundish with wavy, shallow-toothed margins and a red spot at the leaf base.

Mustards *Brassica* spp.

A number of different mustard species are established in California. Most commonly found in orchards are black mustard (*B. nigra*), wild mustard (*B. kaber*) and wild turnip or common yellow mustard (*B. campestris*). Mustards are characterized by rather large lower leaves, usually more or less deeply cut into irregular lobes, and by distinctive seed pods.

E. The mature plant is erect and about 0.5 to 2 meters (2 to 6 feet) tall. The leaves grow alternately on the stem and are toothed or irregularly lobed.

F. All mustards look similar in the seedling stage. The very young seedling may be mistaken for bindweed because both species have broad seed leaves with a deep notch at the tip. The first true leaves, however, are very distinct; they are deeply toothed and often hairy. (Page 86)

G. Many flowers arise at the tip of the stalk. They have four petals, which spread in the form of a cross. The flower stalk extends during the blooming season, leaving maturing seed pods with long beaks. The shape and arrangement of the seed pods are distinctive for different mustard species. (Page 86)

A

Summer Annuals, Grasses

Barnyardgrass *Echinochloa crus-galli*

Several varieties of barnyardgrass or watergrass grow in California; they vary mainly in the appearance of the flower head. Barnyardgrass is the only major summer weedy grass without a ligule.

A. The mature plant is stout, upright growing and varies in height from 15 cm to about 2 meters (6 inches to 6 feet). It often roots at the lower nodes, forming dense tufts. The leaf sheath is flattened and often reddish tinged at the soil line. The lower spikes of the flower head are spaced apart; the top ones are crowded together. The flower groups are barbed with short, stiff hairs or long bristles, depending on the variety.

B. The collar region has no ligule and no auricles.

C. A related species, junglerice (*E. colonum*) is also without a ligule but usually has reddish stripes across the leaf blades.

B

C

Bearded Sprangletop *Leptochloa fascicularis*

Bearded sprangletop is a summer annual commonly found in wet places throughout the interior valleys and is usually associated with alkaline soils.

D. The mature plant is coarse, upright growing and forms large tufts 30 cm to 1 meter (12 to 40 inches) tall. The flower heads are long, erect, well branched and straw colored at maturity.

E. The collar region has a thin, long and delicate ligule and no auricles.

D

E

A

B

C

D

E

F

Summer Annuals, Broadleaf

Prostrate Knotweed *Polygonum aviculare*

Prostrate knotweed is troublesome in orchards because its tough, wiry stems can get entangled in the sweeping machinery at harvest. It thrives even on poor and compacted soil. Knotweed belongs to the buckwheat family; plants in this family have a torn sheath encircling the stem at the leaf base and usually swollen nodes.

A. The mature plant grows prostrate to erect with slender, tough stems, which are extensively branched, giving it a zigzag appearance.

B. The tiny, green flowers with white or pink margins sit in clusters of 2 to 5 on short stalks in leaf axils.

C. The seed leaves are long, very narrow, rounded at the tip and light green with a white cast. The true leaves are much broader, emerging from an encircling, membranous sheath at the leaf base.

Puncturevine *Tribulus terrestris*

Puncturevine has a dense cover of hairs and an extensive root system, which makes it very drought resistant. Its characteristic spiny seed burrs are easily disseminated by animal feet and vehicle tires.

D. The mature plant grows prostrate on open ground but almost erect in dense vegetation. The yellow flowers are borne singly in the leaf axils and open only on sunny mornings, except in shady areas. The fruit consists of a cluster of 5 spiny nutlets or burrs; it breaks apart at maturity.

E. The seedling has thick, elongate and brittle seed leaves, which are green above, grayish underneath and creased along the prominent midvein. The true leaves consist of 8 to 12 leaflets.

Common Purslane *Portulaca oleracea*

Purslane thrives under dry conditions but also competes well with many irrigated crops. It prefers loose, nutrient-rich, sandy soil. Horse purslane (*Trianthema portulacastrum*) looks similar to common purslane but has broader leaves and purple flowers.

F. The mature plant is prostrate or ascending, 15 to 30 cm (6 to 12 inches) tall. The plant branches at the base and along the stems. Leaves are very succulent and often reddish tinged. The small flowers are borne singly or in clusters of 2 to 3 in stem axils or at the tip of stems. They usually open only on sunny mornings. Purslane is often attacked by the purslane sawfly; its larvae feed inside the leaves.

Lambsquarters and Goosefoot *Chenopodium* spp.

Two *Chenopodium* species are often found in orchards, common lambsquarters (*C. album*) and nettleleaf goosefoot (*C. murale*). They grow in waste places, along roadsides and ditches and may germinate from early spring through fall. The mealiness of young leaves, stems and flowers is characteristic of the *Chenopodium* species.

A. The mature plant of common lambsquarters has one main stem, is often heavily branched, grows up to 1.5 meters (5 feet) tall and has a strong tap root. The flowers are arranged in tight clusters at the tip of stems.

B. The seed leaves of common lambsquarters are long and narrow, stalked, dull green and mealy above. The first true leaves are wedge shaped with entire margins and grow opposite on the stem; later leaves are coarsely toothed or lobed and grow alternately.

C. Nettleleaf goosefoot (*C. murale*) resembles common lambsquarters, but the leaves are dark green, fleshy and less mealy, and the flower clusters are borne at the tip of the stems or on short stalks in leaf axils.

A

B

C

Pigweeds *Amaranthus* spp.

Of the several pigweed species growing in California, redroot pigweed (*A. retroflexus*) and tumble pigweed (*A. albus*) are most often found in orchards. The seedlings look similar to the *Chenopodium* species, but the true leaves are not mealy.

D. The mature plant of redroot pigweed (left) is erect, up to 2 meters (7 feet) tall, and usually has one tall main stem and shorter branches arising mostly from the base. The leaves have long stalks and prominent veins on the under side. The flowers are arranged in thick spikes at the tip of the stems or in small clusters in leaf axils. The pinkish to red tap root gave the plant its name. Tumble pigweed (right) can be distinguished from redroot pigweed by its roundish shape, smaller leaves, and small flower clusters borne in leaf axils along the stems. The roots are spreading and relatively shallow. At maturity, the plant breaks away from the roots, becoming a "tumbleweed."

E. The seedlings of pigweeds look very similar. The seed leaves are long oval, bright green above and magenta below. The first true leaves are broadly oval, dark green with entire, wavy margins.

D

E

A

B

C

D

E

F

G

Perennials

Curly Dock *Rumex crispus*

Curly dock is one of the most common docks in California. The docks are members of the buckwheat family; these plants are characterized by jointed stems, a membranous sheath at the leaf base and usually swollen nodes.

A. The mature plant has stout stems, which are often reddish. It may be 0.5 to 1.5 meters (2 to 5 feet) tall or low growing when mowed. The long, loosely branched flower head is green but turns brownish red at maturity. In the fall, the stem dies back and a basal rosette of leaves is formed.

B. Curly dock can regrow from its thick, fleshy taproot when disced or mowed.

Field Bindweed *Convolvulus arvensis*

Field bindweed or morningglory is a troublesome perennial, which reproduces from an extensive rootstock and abundantly produced seeds; the seeds may remain dormant for many years. It prefers dry, warm, nutrient-rich and deep loams.

C. The mature plant grows prostrate with slender stems, 1 meter (3 feet) or longer, or upward while entwining other plants. The flowers are funnel shaped, white to light red and are borne singly on slender stalks in leaf axils. They open only on sunny mornings.

D. The seed leaves are broad and notched at the tip. The leaf stalks are grooved above. The first true leaves are heart shaped and deeply lobed at the base; later leaves are generally shaped like an arrowhead. Young plants beyond the five-leaf stage begin to form their extensive root system and are very difficult to control.

Bermudagrass *Cynodon dactylon*

Bermudagrass is commonly used as turfgrass, but it is troublesome in orchards. It is a perennial that reproduces from rhizomes, stolons and seeds. The root system tolerates salt and drought but is killed when exposed to the sun.

E. The mature plant forms dense mats with spreading and branching stolons. Erect stems arise from the axils of the short leaves on the stolons; these stems are 10 to 45 cm (4 to 18 inches) tall and have longer leaves than the stolons. The flower head consists of 3 to 7 slender spikes radiating from one point at the tip of the stem.

F. The flower head of large crabgrass (*Digitaria sanguinalis*) resembles that of bermudagrass, but the spikes usually arise slightly apart on the stem. As an annual plant, crabgrass does not have stolons or an extensive rootstock. (Page 90)

G. The collar region of bermudagrass has a fringe of short, white hairs and often tufts of long hairs on the margins; auricles are absent. (Page 90)

Dallisgrass *Paspalum dilatatum*

Dallisgrass is occasionally troublesome in orchards under limited tillage and frequently in areas with standing water. It is a bunchgrass and does not produce stolons.

A. The mature plant forms coarse, loose bunches, 30 to 120 cm (1 to 4 feet) high. The flower head consists of 3 to 6 spikes, which are spaced apart and often drooping.

B. The leaf sheath is somewhat flattened; at the base, it is very hairy, often tinged red and usually inflated. The seedling and young plant often grow prostrate.

C. The collar region has a firm, membranous ligule and a few spreading hairs at the margins; auricles are absent.

D. The rhizomes of dallisgrass are fairly short and have short internodes, forming concentric rings.

Johnsongrass *Sorghum halepense*

Johnsongrass is the most troublesome perennial grass in California, growing in annual crops, orchards and along ditchbanks, where it is often spread by irrigation water. Like dallisgrass, it does not form stolons but reproduces from rhizomes and seeds; the seeds may be dormant for many years.

E. The mature plant grows in stout, spreading and very leafy patches, which may be up to 2 meters (6 to 7 feet) tall. The leaves have a prominent whitish midvein, which snaps readily when folded over. The leaf sheath is somewhat flattened and reddish at the soil line but is not hairy, as in dallisgrass. The young plant may resemble barnyardgrass except barnyardgrass does not have a ligule. The flower head is large, open, well branched and often reddish tinged.

F. The ligule consists of a fringe of dense, fine hairs. On the young leaf, the ligule may be membranous with torn margins.

G. The rhizomes are thick, fleshy and white to purplish; they are longer and have longer internodes than those of dallisgrass.

A

B

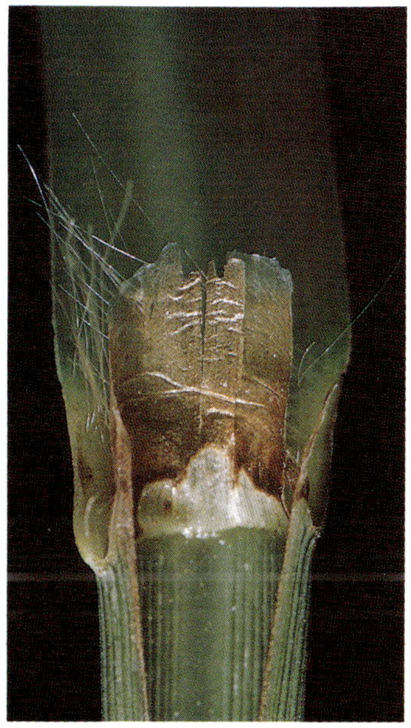

C. ELMORE

C

D

E

F

G

Key to Damage Symptoms

This key or guide briefly describes the major damage and stress symptoms encountered in walnut orchards. It is best to use the key while observing symptoms in the orchard. Look for the category of symptoms (flowers, shoot and leaves, etc.). At each step, you will have two choices; follow the steps until you find the description that matches your observation. Then refer to the individual pest section for detailed descriptions of the pest organisms, the damage symptoms, and the management guidelines.

Because of varying orchard conditions, damage symptoms frequently are not uniform or typical. Furthermore, the apparent damage symptom may be the result of a number of pests or environmental stresses. Many pests or stresses may produce symptoms on different parts of the tree. Disease organisms frequently cause a general decline of the tree in addition to specific symptoms at the site of infection. If symptoms occur on different tree parts, cross references are included in the key. Some of the less common symptoms are not listed. See your farm advisor if you need further help in identifying damage or stress symptoms.

Blight

I. Symptoms on Flowers

black lesions on catkins or pistillate flowers

p. 63 **Blight**
p. 73 **Frost**

San Jose Scale

II. Symptoms on Shoots and Leaves

A. brown to black lesions or spots

1. LEAVES COMPLETELY BLACKENED; ENTIRE TREE AFFECTED

p. 73 **Frost**

2. SMALL SPOTS, MAY COALESCE TO LARGER AREAS; YOUNG LEAVES BECOME DISTORTED; NOT UNIFORM THROUGHOUT TREE

p. 63 **Blight**

3. LEAVES SMALL OR MOTTLED; OR LEAF MARGINS BROWN OR GRAY AND CURLED

p. 73 **Nutritional Disorders**
p. 73 **Windburn**

B. no lesions

1. DIEBACK OF SHOOTS AND/OR TWIGS; SPARSE FOLIAGE, YELLOWING LEAVES

a) scale insects present on bark

p. 50 **Armored Scales**

b) no scale insects present

- trunk, scaffolds or crown affected (see V or VI)
- trunk or scaffold not affected

p. 17 **Water Stress**

2. NO DIEBACK OF SHOOTS AND/OR TWIGS

a) honeydew present, often with sooty mold

- scale insects with frosted or brown cover on twigs

p. 54 **Soft Scales**

- no scale insects present
  - aphids on upper side; midrib blackened

p. 47 **Dusky-veined Aphid**

  - aphids on lower side

p. 47 **Walnut Aphid**

b) no honeydew or sooty mold

- leaves skeletonized
  - no webbing

p. 59 **Redhumped Caterpillar**

  - tentlike webbing

p. 60 **Fall Webworm**

- leaves not skeletonized; light stippling, browning or bronzing
  - mites present
    - — webbing; leaves browning and drop

p. 56 **Two spotted or Pacific Mites**

    - — little or no webbing; leaves bronzed, do not drop

p. 58 **European Red Mite**

  - no mites present; leaves browning and desiccating, do not drop; limited to one or several branches (see also V)

p. 69 **Branch Wilt**

Aphid Honeydew

Husk Fly

III. Symptoms on Husks

A. brown to black lesions; kernels may darken and/or shrivel

1. LESIONS ANYWHERE ON HUSK

   a) lesions firm, sunken, often cracked; no eggs or maggots in husk

   p. 63 **Blight**

   b) lesions soft, not sunken or cracked; eggs or maggots usually in husk

   p. 44 **Husk Fly**

2. LESIONS ON SIDE EXPOSED TO SUN

   p. 72 **Sunburn**

B. no brown or black lesions

1. FRASS EMERGING FROM LARVAL ENTRY HOLE AT BLOSSOM END OR SIDE OF HUSK

   p. 36 **Codling Moth**

2. NO FRASS OR LARVAL ENTRY HOLE; ENTIRE HUSK MAY SHRIVEL

   p. 17 **Water Stress**

Navel Orangeworm

IV. Symptoms on Kernels

A. nutmeat partially consumed; webbing and/or frass

1. LARVA WITH MINUTE BROWN CRESCENT ON BOTH SIDES OF THE SECOND SEGMENT BEHIND HEAD; ONE OR MORE LARVAE PRESENT; PROFUSE WEBBING

   p. 42 **Navel Orangeworm**

2. LARVA WITHOUT CRESCENT ON SECOND SEGMENT; ONE LARVA PRESENT; LITTLE OR NO WEBBING

   p. 36 **Codling Moth**

B. nutmeat not consumed

1. KERNEL DARKENED AND/OR SHRIVELED

   a) husk affected (see III)

   b) husk usually not affected (see also II)

   p. 47 **Aphids**

2. KERNEL WITH FINE THREADS OF FUNGAL GROWTH

   p. 73 **Molds**

Deep Bark Canker

V. Symptoms on Scaffolds or Trunk

A. cankers in bark; ooze

1. LONG, DEEP CRACKS; STREAKS AND PITS OF CANKERED TISSUE IN INNER BARK; SPARSE OR DROOPING FOLIAGE

   p. 70 **Deep Bark Canker**

2. NO CRACKS; GROUPS OF BROWN TO BLACK ROUND SPOTS OR AREAS; CANKER AFFECTS ONLY THE OUTER BARK

   p. 72 **Shallow Bark Canker**

B. no cankers or ooze

1. OUTER LAYER OF BARK PEELING, REVEALING SOOTY MASS OF FUNGAL SPORES; AFFECTED BRANCHES WITH DESICCATED LEAVES THAT REMAIN ON TREE

   p. 69 **Branch Wilt**

2. BARK NOT PEELING BUT WITH SHORT, SHALLOW CRACKS

   a) heavy population of scale insects

   p. 50 **Armored Scales**

   b) no scale insects present

   p. 72 **Sunburn**

J. MIRCETICH

Blackline

VI. Symptoms in Crown Region

A. abnormal outgrowth near ground level

p. 68 **Crown Gall**

B. no outgrowth but cankers present

1. NARROW BLACK STRIP OF DEAD TISSUE UNDER BARK AT GRAFT UNION; OFTEN SUCKERS ON ROOTSTOCK; SPARSE FOLIAGE ESPECIALLY IN THE TREE TOP

   p. 65 **Blackline**

2. NO BLACK STRIP AT GRAFT UNION

   a) whitish fan-shaped fungal plaques under dead, dark brown bark; mushroom odor (see also VII)

   p. 67 **Armillaria**

   b) no fungal plaques under dead, dark brown bark; no mushroom odor when outer bark removed; water-soaked, concentric margins visible between healthy and decayed bark; canker may spread into roots and trunk (see also VII)

   p. 65 **Phytophthora**

VII. Symptoms on Roots

A. entire root system may be affected

1. DECAYING ROOTS FIRM, BRITTLE

   a) shoestringlike threads (rhizomorphs) attached to root (see also VI)

   p. 67 **Armillaria**

   b) no rhizomorphs; cross section of roots reddish brown to black (see also VI)

   p. 65 **Phytophthora**

2. DECAYING ROOTS SOFT, WATERY

   p. 17 **Water Stress**

B. part of root system affected

1. MOSTLY FEEDER ROOTS AFFECTED, STUNTED OR DECAYED; LARGER ROOTS MAY HAVE BLACK LESIONS UNDER SURFACE

   p. 76 **Nematodes**

2. FEEDER ROOTS NOT AFFECTED; LARGER ROOTS WITH GALLS (SEE ALSO VI)

   p. 68 **Crown Gall**

Glossary

aestivation. a state of inactivity during the summer months.

anticoagulant. a substance that prevents the clotting of blood, resulting in internal hemorrhaging; may be used as a rodenticide.

auricles. in grasses, ear-shaped projections at the base of the juncture of leaf sheath and blade (Figure 41).

axil. the upper angle between a twig or leaf and the stem from which it is growing.

biological control. see p. 20.

brood. all the individuals of a generation that hatch at about the same time.

cambium. thin layer of undifferentiated, actively growing tissue between phloem and xylem (Figure 2).

canker. a dead and discolored, often sunken lesion on stem, branch or twig of a plant.

catkin. a spikelike cluster of unisexual flowers, e.g., the male flowers of walnut.

collar region. in grasses, the region where leaf blade and sheath meet (Figure 41).

control action guidelines. see p. 15.

crawler. immature, mobile stage of scale insects.

crown. point at or just below the soil surface where main stem (trunk) and roots join.

cultivar. a cultivated plant variety.

degree-day. a unit combining temperature and time, used to determine growth rates (see also p. 15).

diapause. a period of physiologically enforced dormancy.

dormant spray. treatment applied during the period when trees are inactive.

English walnut. the walnut species (*Juglans regia*) used for the selection of commercial scion cultivars; origin believed in Persia (= Persian walnut).

evapotranspiration. the quantity of water retained in plant tissue, transpired through leaf pores or evaporated from adjacent soil surfaces in a specified time period.

feeder roots. the youngest roots with root hairs, important in absorption of water and minerals.

flight. period of flying activity of moths from one generation.

frass. insect excrement.

gall. localized swelling or outgrowth of plant tissue, often formed in response to the action of a pathogen or other pest.

inner bark. in older trees, the living part of the bark, comprised of phloem and cambium (Figure 2).

instar. the stage of an insect between successive molts.

lesion. localized area of diseased or discolored tissue.

ligule. in many grasses, a short projection on the inner side of the leaf blade where it meets the leaf sheath (Figure 41).

molt. in insects and other arthropods, the shedding of skin before entering another stage of growth.

mummy. unharvested nut remaining on the tree (also called sticktight); the crusty skin of an aphid whose inside has been consumed by a parasite.

mycelium. the vegetative body of a fungus consisting of a mass of long filaments or hyphae.

nymphs. the immature stages of insects that have incomplete metamorphosis (i.e., do not have a pupal stage).

outer bark. in older trees, the dead part of the bark (Figure 2).

packing tissue. firm, membranous tissue lining the walnut shell and separating the kernel halves (Figure 4).

pathogen. any disease-producing organism.

Persian walnut. see English walnut.

pest resurgence. see p. 21.

pesticide resistance. see p. 23.

pheromone. a substance secreted by insects to affect the behavior or development of other members of the species (see also p. 35).

phloem. the food-conducting tissue of vascular plants (Figure 2).

pistillate flower. a female flower.

rhizome. a trailing underground shoot, often rooting at the nodes and growing new plants (Figure 41).

rootstock. lower portion of a graft which develops into the root system.

scion. in a grafted tree, the upper portion of a graft from which the stem and branches grow.

secondary pest outbreak. see p. 22.

seed leaves. the first leaf (grasses) or first two leaves (broadleaf plants) on a seedling (Figure 41).

sessile. attached or fastened, incapable of moving from place to place; attached directly without a stem.

sheath. the part of a grass leaf that encloses the stem below the collar region (Figure 41).

staminate flower. a male flower.

sticktight. nut that remains on the tree after harvest (also called mummy); nut with husk firmly adhering to shell.

stolon. a trailing aboveground shoot, often rooting at the nodes and forming new plants (Figure 41).

sucker. shoot arising at or below ground level.

suture. seal between the two halves of the nut shell.

true leaves. all leaves a plant produces after the seed leaves.

xylem. the vessels conducting water and minerals from the roots to upper plant parts. (Figure 2).